# International and Development Education

The *International and Development Education Series* focuses on the complementary areas of comparative, international, and development education. Books emphasize a number of topics ranging from key international education issues, trends, and reforms to examinations of national education systems, social theories, and development education initiatives. Local, national, regional, and global volumes (single authored and edited collections) constitute the breadth of the series and offer potential contributors a great deal of latitude based on interests and cutting edge research. The series is supported by a strong network of international scholars and devᵉˡ t professionals who serve on the International and Development Education Aᵈ participate in the selection and review process for manuscript devᵉˡ

## SERIES EDITORS
**John N. Hawkins**
*Professor Emeritus, University of California, Los Aᵢ*
*Senior Consultant, IFE 2020 East West Center*

**W. James Jacob**
*Assistant Professor, University of Pittsburgh*
*Director, Institute for International Studies in Education*

## PRODUCTION EDITOR
**Heejin Park**
*Project Associate, Institute for International Studies in Education*

## INTERNATIONAL EDITORIAL ADVISORY BOARD
**Clementina Acedo,** *UNESCO's International Bureau of Education, Switzerland*
**Philip G. Altbach,** *Boston University, USA*
**Carlos E. Blanco,** *Universidad Central de Venezuela*
**Sheng Yao Cheng,** *National Chung Cheng University, Taiwan*
**Ruth Hayhoe,** *University of Toronto, Canada*
**Wanhua Ma,** *Peking University, China*
**Ka-Ho Mok,** *University of Hong Kong, China*
**Christine Musselin,** *Sciences Po, France*
**Yusuf K. Nsubuga,** *Ministry of Education and Sports, Uganda*
**Namgi Park,** *Gwangju National University of Education, Republic of Korea*
**Val D. Rust,** *University of California, Los Angeles, USA*
**Suparno,** *State University of Malang, Indonesia*
**John C. Weidman,** *University of Pittsburgh, USA*
**Husam Zaman,** *Taibah University, Saudi Arabia*

**Institute for International Studies in Education**
School of Education, University of Pittsburgh
5714 Wesley W. Posvar Hall, Pittsburgh, PA 15260 USA

**Center for International and Development Education**
Graduate School of Education & Information Studies, University of California, Los Angeles
Box 951521, Moore Hall, Los Angeles, CA 90095 USA

## Titles:

*Higher Education in Asia/Pacific: Quality and the Public Good*
Edited by Terance W. Bigalke and Deane E. Neubauer

# Higher Education, Policy, and the Global Competition Phenomenon

Edited by

*Laura M. Portnoi, Val D. Rust, and Sylvia S. Bagley*

HIGHER EDUCATION, POLICY, AND THE GLOBAL COMPETITION PHENOMENON

Copyright © Laura M. Portnoi, Val D. Rust, and Sylvia S. Bagley, 2010.

All rights reserved.

First published in hardcover in 2010 by PALGRAVE MACMILLAN® in the United States—a division of St. Martin's Press LLC, 175 Fifth Avenue, New York, NY 10010.

Where this book is distributed in the UK, Europe and the rest of the world, this is by Palgrave Macmillan, a division of Macmillan Publishers Limited, registered in England, company number 785998, of Houndmills, Basingstoke, Hampshire RG21 6XS.

Palgrave Macmillan is the global academic imprint of the above companies and has companies and representatives throughout the world.

Palgrave® and Macmillan® are registered trademarks in the United States, the United Kingdom, Europe and other countries.

ISBN: 978–1–137–36655–9

Library of Congress Cataloging-in-Publication Data is available from the Library of Congress.

A catalogue record of the book is available from the British Library.

Design by Newgen Knowledge Works (P) Ltd., Chennai, India.

First PALGRAVE MACMILLAN paperback edition: November 2013

10 9 8 7 6 5 4 3 2 1

*We gratefully dedicate this book to our spouses—Howard, Diane, and Ross—whose support throughout the process of this endeavor has been invaluable.*

# Contents

# Tables and Figures

## Tables

## Figures

# Acknowledgments

The editors would like to thank several people for their assistance with the production of this volume. First, all of the contributing authors have been wonderful in embracing the project and instrumental in completing their work under a tight production schedule. Second, Marijana Benesh provided specialized editing assistance, as did our series editors, W. James Jacob and John N. Hawkins. We appreciate the series editors' guidance throughout the entire process. Finally, we would like to thank Julia Cohen, Burke Gerstenschlager, and Samantha Hasey at Palgrave Macmillan for their support with the project.

# Acronyms and Abbreviations

| | |
|---|---|
| ACA | Academic Cooperation Association |
| ACE | American Council on Education |
| ADIAC | Australia's Department of Immigration and Citizenship |
| AEI | Australian Education International |
| AHELO | Assessment of Higher Education Learning Outcomes |
| AIEA | Association of International Education Administrators |
| AIU | Asian International University |
| ANQAHE | Arab Network for Quality Assurance in Higher Education |
| APQN | Asian-Pacific Quality Network |
| ARWU | Academic Ranking of World Universities |
| ASEAN | Association of Southeast Asian Nations |
| AUIDF | Australian Universities Directors Forum |
| AUN | Asian University Network |
| AUQA | Australian Universities Quality Agency |
| BNU | Beijing Normal University |
| CAA | Commission for Academic Accreditation |
| CANQATE | Caribbean Area Network for Quality Assurance in Tertiary Education |
| CEE | Network of Central and Eastern Europe Quality Assurance Agencies in Higher Education |
| CHED | Center for Higher Education Development |
| CHEDC | Center for Higher Education Development Chair |
| CIA | Central Intelligence Agency |
| CIEE | Council on International Educational Exchange |
| CMOE | China's Ministry of Education |
| CONEAU | Comisión Nacional de Evaluación y Acreditación Universitaria |
| CONICET | Consejo Nacional de Investigaciones Científicas y Técnicas |
| CREATE | Campus for Research Excellence and Technological Enterprise |
| CWTS | Center for Science and Technology Studies |

| DAAD | Deutscher Akademisher Austausch Dienst |
|---|---|
| EAIE | European Association for International Education |
| EGC | Education for Global Competition |
| EGM | Emerging Global Model |
| ENQA | European Association for Quality Assurance in Higher Education |
| ERASMUS | European Action Scheme for the Mobility of University Students |
| ESI | Essential Science Indicators |
| ESOS | Educational Services for Overseas Students (Australia) |
| EUA | European University Association |
| FDI | Foreign Direct Investment |
| FOMEC | Fondo para el Mejoramiento de la Calidad Universitaria |
| GATE | Global Alliance for Transnational Education |
| GATS | General Agreement on Trade and Services |
| GDP | Gross Domestic Product |
| GCC | Gulf Cooperation Council |
| HEIs | Higher Education Institutions |
| HEQC | Higher Education Quality Committee (South Africa) |
| IAU | International Association of Universities |
| ICDE | International Council for Open and Distance Education |
| IDP | International Development Program (Australia) |
| IFIs | International Financial Institutions |
| IIE | Institute of International Education |
| IIMD | International Institute for Management Development |
| IITs | Indian Institutes of Technology |
| IMHE | Institutional Management in Higher Education |
| INQAAHE | International Network for Quality Assurance Agency in Higher Education |
| IMF | International Monetary Fund |
| IPRs | Intellectual Property Rights |
| INSEAD | Institut Européen d'Administration des Affaires |
| IREX | International Research and Exchanges Board |
| ISI | Institute for Scientific Information |
| IT | Information Technology |
| JKUAT | Jomo Kenyatta University of Agriculture and Technology |
| KAUST | King Abdullah University of Science and Technology |
| KMOE | Kyrgyz Ministry of Education |
| KU | Kenyatta University |
| MESDCs | Major English Speaking Destination Countries |
| MFN | Most Favored Nation |
| MIT | Massachusetts Institute of Technology |

MINCYT    Ministerio de Ciencia, Tecnología e Innovación Productiva
          (Ministry of Science, Technology and Innovation in
          Production, Argentina)
MMOHE     Malaysia's Ministry of Higher Education
MQF       Malaysian Qualifications Framework
NAFSA     Association of International Educators
NOTCFL    National Office for Teaching Chinese as a Foreign Language
NECTA     National Examinations Council (Tanzania)
NRF       National Research Foundation (South Africa)
OCHE      Oman's Council on Higher Education
OECD      Organization of Economic Cooperation and Development
OHEAQ     Oman Higher Education Access Questionnaire
OMOF      Oman's Ministry of Finance
OMOHE     Oman's Ministry of Higher Education
OMOM      Oman's Ministry of Manpower
OMONE     Oman's Ministry of National Economy
PISA      Programme for International Student Assessment
PKU       Peking University
PPP       Purchasing Power Parity
R&D       Research and Development
RIACES    *La Red Iberoamericana para la Acreditación de la Calidad de
          la Educación Superior*
RMB       Renminbi
RO        Omani riyals
SAPs      Structural Adjustment Programs
SEDP      Secondary Education Development Plan (Tanzania)
SCI       Science Citation Index
SCIE      Science Citation Index-Expanded
SCU       Sichuan University
SCUP      Society for College and University Planning
SJTUIHE   Shanghai Jiao Tong University Institute of Higher
          Education
SQU       Sultan Qaboos University
SSCI      Social Sciences Citation Index
STEM      Science, Technology, Engineering, and Mathematics
TAFE      Technical and Further Education (Australia)
THE       *Times Higher Education*
THU       Tsinghua University
TJU       Tianjin University
TMEC      Tanzania's Ministry of Education and Culture
TMHEST    Tanzania's Ministry of Higher Education, Science, and
          Technology

| | |
|---|---|
| TMEVT | Tanzania's Ministry of Education and Vocational Training |
| TTIE | Tanzania's Institute of Education |
| UAE | The United Arab Emirates |
| UAEMOE | United Arab Emirates' Ministry of Education |
| UAEMHESR | United Arab Emirates' Ministry of Higher Education and Scientific Research |
| UAEU | United Arab Emirates University |
| UBA | *Universidad de Buenos Aires* |
| UCB | University of California, Berkeley |
| UCT | University of Cape Town |
| UDSM | University of Dar es Salaam |
| UGA | University of Georgia |
| UM | University of Michigan, Ann Arbor |
| UMAP | University Mobility in the Asia Pacific |
| UNAM | *Universidad Nacional Autonoma de Mexico* |
| UNCTAD | United Nations Conference on Trade and Development |
| UNESCO | United Nations Educational, Scientific and Cultural Organization |
| UNICEF | United Nations Children's Fund |
| UNISA | University of South Africa |
| UofN | University of Nairobi |
| USAID | United States Agency for International Development |
| USIA | United States Information Agency |
| VMOET | Vietnam's Ministry of Education and Training |

# Series Editors' Introduction

During the inaugural Higher Education Special Interest Group (HESIG) business meeting of the Comparative and International Education Society (CIES) Annual Meeting in New York in 2008, HESIG leaders launched a HESIG-sponsored book project on current topics in international higher education. This is the first volume published in the *International and Development Education Book Series* co-sponsored by HESIG. Contributors to this volume were solicited from an initial Call for Chapter Proposals distributed initially to HESIG members and afterward to all CIES members.

Higher education institutions are increasingly required to compete with each other on several levels and often adjust their recruitment, curricula, and missions in accordance to the market needs. Competition is often viewed as a conducive way to improve the quality of a nation's higher education system. Some economists argue that higher education has the potential to stimulate economic growth, especially in emerging markets such as China, India, and Brazil.

Editors Laura M. Portnoi, Val D. Rust, and Sylvia S. Bagley come uniquely qualified to introduce an overview debate on the topics reflected in this latest volume's title *Higher Education, Policy, and the Global Competition Phenomenon*. Portnoi is an expert on higher education policy, with extensive experience in the United States and South Africa. In addition to serving as a former president of CIES, Rust has published widely on higher education issues at the global, national, and local levels and has headed research and technical assistance projects in Azerbaijan, Ethiopia, Georgia, Germany, India, Norway, Pakistan, Senegal, and the United States. Bagley has served in a number of higher education administrative positions directing international education projects and study abroad programs.

The volume represents a comprehensive discussion of the trend toward competition in higher education across the globe, as well as the complex dimensions of the competition phenomenon. The editors provide both an overview and analysis of the competition phenomenon from the global landscape as well as bring together a group of country case study chapters. The historical development of global rankings of higher education institutions is discussed in several chapters as an area of recognition and debate as well as competition. Market forces and their impervious link to the higher education subsector are introduced as well as the inevitable negative repercussions that are often associated with higher education forged through a market focus. The emergence of a knowledge economy is also discussed by several contributors, and is the focus of Simon Marginson's chapter 3 on the *k-economy*. National case study chapters include an examination of higher education in Argentina, Australia, China, Malaysia, Oman, Tanzania, and Vietnam.

This book will be an excellent resource for university faculty members and students concerned with higher education in general and especially with comparative and international interests. It will also be of benefit to policy makers and planners, higher education specialists, governmental officials, and senior administrators in the higher education subsector.

John N. Hawkins
*University of California, Los Angeles*

W. James Jacob
*University of Pittsburgh*

# Chapter 1

# Mapping the Terrain: The Global Competition Phenomenon in Higher Education

*Laura M. Portnoi, Sylvia S. Bagley, and Val D. Rust*

In an increasingly open and integrated world, competition has become a central preoccupation of the higher education subsector. With the collaboration of global financial institutions and development agencies, neoliberalism—an economic paradigm that favors a free-market economy and decreased government involvement—has swept the world in recent decades. The competitive forces that accompany the global free-market economic system impact every level of the higher education subsector in significant ways. Combined with the impact of globalization and the development of the worldwide "knowledge economy," these competitive forces have resulted in the *global competition phenomenon* that is currently reshaping the higher education subsector.

Several developments characterize the global competition phenomenon. One of its key features is the manifestation of academic capitalism, distinguished by universities as entrepreneurial marketers and knowledge as a commodity rather than a public good (Slaughter and Rhoades 2004). Another facet of the phenomenon is an increase in institutional mergers, which may involve melding various configurations of "strong" and "weak" institutions to enhance the universities' or a country's competitive advantage (Harman and Harman 2008). With growing demand in the free-market system, the global higher education environment is also experiencing increased private and cross-border provision of instruction, as well as student mobility. The advent of the global knowledge economy, in which knowledge is produced and exchanged, is also a key development that is highly intertwined with the movement toward competition.

In the evolving hierarchical and unequal global higher education system, being competitive becomes key, and global positioning is integral to competing with other nations and institutions (Marginson 2006). As Frans van Vught (2008) points out, universities are currently in a "reputation race," in which they compete for reputation and academic prestige. Furthermore, Marginson (2006) argues that "the more an individual university aspires to the top end of competition, the more significant global referencing becomes" (27). Universities, and the countries in which they are located, thus seek to project the best image possible in order to be poised to compete for research funding, the "best and brightest" international students, and "star" faculty members.

Yet, despite discernable global trends, the parameters of the global competition phenomenon remain fluid, and the developments taking place are multifaceted. For example, while neoliberal economic practices have resulted in government divestment in higher education, due to national and regional dynamics many countries in Asia and the Middle East have increased their investment in higher education in recent years. However, much of this investment is geared toward the development of private institutions or "world-class" research universities that serve a specific subset of the population, raising questions regarding access and equity.

Competition between and among universities takes on multiple forms and may occur at the institutional, local, regional, national, and global levels. For instance, cross-border provision of higher education involves competition between domestic providers, which may be public or private, and foreign institutions that conduct business in the local environment. Furthermore, as evidenced by the chapters in this volume, there are no central change agents in the global competition phenomenon—developments may be impacted or instituted by international aid agencies and financial institutions, national governments, individual institutions, or administrators. In some cases governments that seek to improve their competitive advantage invest in the development of specific institutions, as with "world-class" universities, and in others, individual institutions employ their own strategies relevant to their context. External international bodies may also initiate changes, as is the case with global university rankings conducted by specific organizations.

The emergence of global university ranking systems, most notably Shanghai Jiao Tong University Institute of Higher Education's *Academic Ranking of World Universities*, also known as the Shanghai index or Shanghai rankings, and the *Times Higher Education* (*THE*) World

University Rankings, is a significant development. Though a number of countries have internal ranking mechanisms, or league tables, ranking institutions across borders brings competition to the global level. The *THE* composite rankings involve reviews by academics and employers, as well as university indicators, such as faculty-student ratio and the number of citations per faculty member. The Shanghai rankings rate institutions on four "quality" measures, specifically related to research: (a) quality of education: the number of Nobel Prizes and Fields Medals awarded to alumni; (b) quality of faculty: the number of faculty awarded Nobel Prizes and Fields Medals or listed as "highly cited" researchers in 21 categories; (c) research output: papers published in *Science* and *Nature* and articles published in the Science Citation Index-Expanded and Social Sciences Citation Index; (d) per capita performance: weighted scores of the first three measures are added and divided by the number of faculty members (Labi 2008).

These ranking mechanisms are a by-product of the competition phenomenon; at the same time they engender increased competition as universities clamor to make it to the top of the list or to be represented at all. Furthermore, higher education leaders are increasingly using these rankings to make decisions and to influence reform (Hazelkorn 2008). However, the validity (and value) of these rankings are questionable. Ranking mechanisms are created with specific sets of indicators designed to represent quality; thus, some indicators are omitted, and various ranking mechanisms focus on different indicators (Usher and Savino 2006). Alex Usher and Massimo Savino (2006) point out that what institution is "best" depends on the indicators and weightings chosen by the publisher. Thus, each ranking system implicitly defines educational quality through the indicators selected and the distribution of its weighting mechanisms.

Moreover, "all of this emphasis to performances gravitates towards an ideal, a typical picture of a particular type of institution," (Huisman 2008, 2), what Kathryn Mohrman, Wanhua Ma, and David Baker (2008) call the Emerging Global Model (EGM) of the top stratum of research universities. Rankings' emphasis on research and traditional academic performance leads to "mimicking" behavior, in which high reputation research institutions are imitated (van Vught 2008). Rankings may therefore lead institutions to focus on what is, for many, an unobtainable goal— becoming a "world-class" research university—at the expense of other university missions, such as teaching and service to society. Furthermore, the EGM threatens the diversity of higher education institutions (HEIs); many types of institutions are needed to serve growing demand and multiple local, national, and global purposes (Huisman 2008; van Vught 2008; Marginson, chapter 3). The intricacies of the movement toward rankings and its implications for global competition are thus complex and impact various institutional types in differing ways.

Given the divergences and variation involved, is competition in higher education truly a global phenomenon? In Bob Lingard's (2000) words, "Well, it is and it isn't" (79). There are indeed common trends occurring around the world within higher education, many of which are related to increasing competition between countries and individual institutions. At the same time, however, strategies and developments related to competition take on a local character—regionally, nationally, and at individual institutions. Moreover, the relationship between the knowledge economy, market or market-like forces, commercialization, and international rankings is multidirectional in the current context of global competition.

The chapters in this volume provide rich evidence of the complexities of the global competition phenomenon in higher education. The authors investigate multiple aspects of this trend and discuss many of the tensions involved. Some authors focus on the broad changes impacting the sector, while others discuss the varying ways global competition plays out at the local level. Furthermore, the authors encourage debate by covering the various facets involved in both similar and different ways. Despite the variety of topics and angles presented, several interconnected themes permeate throughout the volume: (a) the knowledge economy and the rise of global ranking mechanisms, (b) varying actors, varying purposes, and strategies, (c) quality assurance, and (d) internationalization and student mobility.

## The Knowledge Economy and the Rise of Global Ranking Mechanisms

Setting the stage for the volume, Lynn Ilon (chapter 2) provides an overview of global economic changes over the past 50 years that have resulted in the merging of the first, second, and third worlds. She outlines "the global economic environment into which higher education has been thrust" (15), noting that higher education has been continually moving out of the realm of the government and into the private sector. Thus, as Jane Knight points out in chapter 15, higher education is increasingly characterized by commercialization and commodification due to pervasive market or market-like forces. Furthermore, Diane E. Oliver and Nguyen Kim Dung (chapter 10), Robin Shields and Rebecca Edwards (chapter 17), and Knight (chapter 15) all discuss higher education "officially" becoming a tradable commodity in many countries as a result of the provisions of the General Agreement on Trade and Services (GATS).

As Simon Marginson points out in chapter 3, the emergence of the worldwide knowledge economy, involving the codification and valuation of knowledge, has significance for the changing nature of higher education in the current era. He argues further that the emergence of global ranking mechanisms is closely synchronized to worldwide developments in the knowledge economy. This view regarding the connection between the global knowledge economy, increasing competition, and the focus on rankings resonates with several other chapters in this volume. For Francisco O. Ramirez (chapter 4), however, the key reason universities are ranked across borders is the worldwide shift from the university as a national, historical institution to the university as a rationalized organization.

Whatever the purpose, the use of global ranking mechanisms raises numerous concerns. In chapter 5, Isaac Ntshoe and Moeketsi Letseka contend that rankings are imposed through policy borrowing and lending by more established countries onto less developed ones. The authors also question the measures on which the rankings are based. Furthermore, like Kathryn Mohrman and Yingjie Wang (chapter 12), Ntshoe and Letseka point out that rankings often highlight the sciences at the expense of social sciences and humanities. Moreover, Marginson posits that while the Shanghai index does capture "real" measures such as research output, rather than simply reputation, the list is often read as a "world's best list," not a research performance list. Similarly, Ntshoe and Letseka argue that given the lack of internationally accepted quality assurance mechanisms, rankings are used as a proxy for quality without taking context into account. Furthermore, both Marginson and Anthony Welch (chapter 11) point out that such rankings are biased toward the English-speaking world, because the indicators used for publications are English-language. Due to all these factors, the trend toward rankings—and the value placed upon them—effectively translate into certain countries and institutions remaining on the "knowledge periphery," as Welch contends is the case with Vietnam and Malaysia.

## Varying Actors, Varying Purposes, and Strategies

As noted above, several different types of actors are involved in global competition, from HEIs to governments to organizations that create rankings. Their involvement in increasing the competitiveness of their institutions (and countries) has different purposes, depending on the context. Furthermore, global competition involves a unique policy arena, in which many of the changes employed are not necessarily official policies, but rather *strategies* used by the various entities involved. As we shall see below, some initiatives are indeed official policies, such as China's efforts to create "world class" institutions. Yet, many strategies that various actors of the global competition phenomenon utilize are not as prescriptive, and there are often no clear drivers of such initiatives. This hands-off, unregulated approach is a facet of free-market economics that has encouraged competition on many levels in the higher education subsector.

In the new economy, government involvement may vary depending on the local context, development needs, and the wealth of each country. For example, Gerald Wangege-Ouma (chapter 8) demonstrates how in Kenya, declining government revenues has led HEIs to generate income by operating dual-track programs in which fee-paying students take courses in programs that run parallel to those of government-subsidized students. While government funding (and involvement) has decreased in Kenya and in many other parts of the world, the governments of the United Arab Emirates (UAE) and Oman have been expanding higher education (see chapter 7 and chapter 9). Another type of governmental funding for higher education involves focusing funding and efforts on developing a small number of "world-class" institutions that can complete with major world players, as is the case in China (see Mohrman and Wang, chapter 12).

In Oman and the UAE, the expansion of higher education impacts not only public institutions, but also private forms of higher education. In the UAE, the government has chosen to import higher education programs and personnel from other countries, thus evolving into a "hybrid system, in which rapid private provision is being encouraged and supported by governmental initiatives alongside the more modest expansion of federal institutions" (Kirk and Napier, chapter 9, 116). Likewise, in Oman, the government has fostered a similar dual-provision system in which the expansion of private cross-border HEIs has been encouraged through providing land, tax incentives, and capital grants to private providers (Ameen et al., chapter 7). According to Welch (chapter 11), a similar situation involving a combination of private and public provision of higher education is occurring in Southeast Asian countries, such as Vietnam and Malaysia, due to the rapid expansion of higher education and the dual pressures of rapidly increasing enrollments and declining funding. Thus, one of the specific strategies governments and institutions have employed has been expansion in the private sector as a response to unmet demands. In essence, a hybrid system has evolved between public and private provision of higher education in Oman, Kenya, Vietnam, Malaysia, and the UAE, though all have taken different forms and government involvement has varied.

Differences also exist regarding the purpose of higher education expansion, based on the local context. For example, the UAE and Oman had few

HEIs prior to embarking on expansion agendas. Daniel Kirk and Diane Brook Napier (chapter 9) argue that the UAE's dramatic expansion in higher education is connected both to national development and to positioning itself regionally and globally. University education is thus linked to human resource needs. Likewise, the Tanzanian government has also been focusing on increasing economic development, and has instituted educational reforms that impact higher education, backed in part by the World Bank and other development organizations (chapter 13). Such developments are supported by what Ashley Shuyler and Frances Vavrus call the "discourse of education for global competition" (177). Similarly, Welch (chapter 11) contends that following the directives of international institutions such as the World Bank and the Asian Development Bank, expansion of higher education in Southeast Asia is focused on social and economic development, given that "higher education is the key production site for the highly skilled personnel who, in a post-Fordist world, are believed to be the foundation of the new knowledge economies" (145). Reflecting this notion, the Omani government has been investing in higher education with the goal that its new competitive advantage in the global marketplace might be to provide high quality workers to other countries, given that oil revenues are expected to decline in the near future (Ameen et al., chapter 7). This move could be seen as an ultimate form of higher education, or its "product" (graduates), becoming a commodity. In contrast, other motives are at work in China. Here, the government is focused on a small number of institutions, thus educating the elite (Mohrman and Wang, chapter 12). Mohrman and Wang argue that through Projects 985 and 211, the Chinese government has invested considerable funds in a limited number of choice institutions in order to become internationally competitive.

Nations and individual institutions have enacted further strategies to gain a competitive advantage. In chapter 14, Joseph Stetar et al. discuss different approaches to the use of soft power, which "draws on the subtle effects of culture, values and ideas, in contrast to the more direct, tangible measures" of hard power (192). The authors argue that the United States and China are using soft power to gain global advantage by spreading their cultures around the world. In addition, according to Stetar et al., both countries are using their soft power to recruit international university students—an idea that resonates with Shields and Edwards' discussion in chapter 17. Meanwhile, Stetar et al. note that in Azerbaijan and Kyrgyzstan, various religious, cultural, and linguistic groups are vying for "soft power" interest in their own universities. Based in specific local contexts, this discussion provides an interesting counterpart to Hans de Wit and Tony Adams' analysis in which they document the shift from "aid to trade" in Europe and Australia (chapter 16). In the current higher education climate, Ramirez posits in chapter 4 that another strategy is at work. Writing on the pursuit of excellence in higher education, he argues that universities are responding to global competition by formalizing faculty assessments and standardizing university assessments. Shuyler and Vavrus (chapter 13) explain how the Tanzanian government is employing yet another strategy, seeking to make the country more internationally competitive by replacing the traditional pedagogical mode with student-centered, social constructivist pedagogy. However, based on their empirical research, the authors raise questions about the efficacy and quality of such programs, and, thus their effectiveness in increasing the country's stature globally.

# Quality Assurance

Changes within the higher education system, including increased emphasis on accountability and the advent of global ranking indexes, are also implicated in the movement toward instituting quality assurance mechanisms in countries around the world. However, whether or not countries have implemented their own quality assurance systems, ensuring quality is a complex and contested task. In chapter 5, Ntshoe and Letseka provide an overview of the quality assurance movement, including a brief discussion of various global and regional quality assurance mechanisms. Echoing the discussions of commercialization by Knight (chapter 15) and Ilon (chapter 2), they discuss how notions of quality and quality assurance are shaped by global competition, internationalization, and the discourse of new managerialism, through which a business ethos and drive for accountability has entered higher education. In line with these notions, Wangege-Ouma (chapter 8) contends that parallel programs in Kenya are consistent with a neoliberal, free market ideology in which higher education has become a tradable commodity. Wangege-Ouma argues further that the Kenyan government's pursuit of "economic self determination" through dual track programs leads to goal displacement, in which quality takes a backseat to economic efficiency.

Providing an example of how the trend toward quality assurance plays out at the local level, Héctor Gertel and Alejandro Jacobo (chapter 6) focus on the alignment of Argentine universities to the global quality assurance movement and explore the reasons why the Argentine reaction has been slow. With historical background and analysis of the current context, they provide evidence for their hypothesis that "a slow response results from the complexity associated with the decision-making process of collegiate governing bodies" (95). They contend that the benchmarks set up by new quality assurance standards involve a social, as well as a technical, dimension and argue further that the institution of a quality assurance mechanism has been slow due to the way the autonomous governing bodies function as social units.

While quality assurance mechanisms have become common in many countries around the world, most only evaluate national universities, leaving cross-border education unregulated. Several chapters in this volume provide evidence for the complex ways in which educational expansion, particularly of private and cross-border higher education, may impact quality and access. For instance, Hana Ameen, David Chapman, and Thuwayba Al-Barwani's research in Oman (chapter 7) indicates that educators are concerned about the quality of higher education, particularly private higher education providers. Interestingly, a substantial numbers of the senior-level educators, government officials, and private sector employees who participated in their study considered it to be the Omani government's responsibility to raise the quality of higher education. Oman requires all private HEIs to have an affiliation with an international counterpart to improve quality; however, no specific mechanisms regulating or ensuring quality have been established. Furthermore, Ameen et al.'s findings highlight the tension between quality and access due to the inverse relationship between increasing access (particularly to private institutions) and quality. Questions of access and equity thus arise in this context, as in the UAE, regarding which populations are being served by educational expansion.

Similarly, in Malaysia and Vietnam, concerns about quality pervade, given the arrival of cross-border private higher education providers. Welch (chapter 11) points out that Malaysia has a new qualifications authority, while quality control is more difficult in Vietnam. Furthermore, he argues that there are problems with cronyism in both Malaysia and Vietnam, and corruption in Vietnam. Ethnic discrimination is also a persistent problem in Malaysia, and access is also an issue in Vietnam, where many more students sit for the national exams than can be placed in public institutions (Oliver and Nguyen, chapter 10). Furthermore, both Welch (chapter 11) and Oliver and Nguyen (chapter 10) argue for more regulation of the private, cross-border higher education industry in Vietnam. Yet, Oliver and Nguyen acknowledge the struggle between ensuring careful regulation of cross-border educational services, while also providing an environment that is conducive for the expansion of cross-border education. Even if regulatory measures are in place, however, they may become too rigid to allow flexible reaction to globalization (Gertel and Jacobo, chapter 6). Furthermore, in chapter 5 Ntshoe and Letseka raise a concern about the potential loss of regional and national sovereignty and the disregard for context in the international quality assurance movement. They also raise the question: "[Can we] assume that once mechanisms, processes, criteria, or standards are clearly defined and described, quality in higher education will be assured?" (62).

# Internationalization and Student Mobility

The trend toward global competition also impacts, and is impacted by, the internationalization of HEIs. Furthermore, as Knight points out in chapter 15, internationalization is both a reacting force and an agent of globalization. Through an analysis of the opinions of senior academics and administrators, Knight examines shifting rationales driving processes aimed at developing the international dimension of higher education. Her focus is on the growing importance and impact of competition and the commercial agenda. She posits that the rationale for internationalization is moving from an historical emphasis on academic and cultural exchange to economic and political motivations. She also raises the question of whether increased competitiveness leads to more benefits or risks for higher educa- tion, and calls for increased attention to the complex interplay between cooperation and competition in the current context.

In chapter 16, de Wit and Adams further investigate the trends Knight discusses. Similar to Knight's research, their evaluation of international- ization in Australia and Europe illustrates a shift toward competition over cooperation. The authors focus primarily on policies related to interna- tional students, and demonstrate how by the mid-1980s, Australia and the United Kingdom had moved from "aid to trade" by increasingly requir- ing international students to pay fees, while this shift has been less dra- matic and swift in continental Europe. De Wit and Adams note that more recently, global competition for highly skilled workers has become a strong pull factor in international student circulation due to the "graying" soci- eties of Europe that need to fill the gaps in their knowledge economies. According to the authors, migration and circulation of these individuals, as well as global competition for talent, are becoming commonplace.

In light of this trend toward increased student mobility, Shields and Edwards (chapter 17) discuss the recruitment strategies of well-established universities and those in emerging hubs, such as Saudi Arabia and Singapore. Echoing Welch's (chapter 11) point regarding Malaysia as a new player in attracting international students, they explore how these emerg- ing hubs are drawing increasing numbers of students and thus impacting patterns of international student mobility and competition in higher education. At the same time, well-established universities are taking measures to protect their global preeminence. The authors' analysis suggests that "the degree to which most universities are competitive in recruiting international students will increasingly be shaped by the scope and character of their global network rather than any intrinsic characteristics of their individual courses, culture, or campuses" (236). Internationalization thus presents complex new trends in the higher education subsector, with competition for international students becoming another strategy for seeking competitive advantage. Yet, as Stetar et al. point out in chapter 14, the relationship between competition and cooperation is intricate, and some governments or institutions may even use "soft power" strategies, such as celebrating a country's culture, language, and religion overseas to gain a competitive advantage. Therefore, the distinction between cooperation and competition has become blurred.

# The Complex Terrain of the Global Competition Phenomenon

Taken together, the chapters in this volume provide evidence of the trend toward global competition in higher education and its accompanying ramifications. However, many of the authors' analyses also point to variation in how the competition phenomenon impacts, and is impacted by, global, local, and institutional actors across the worldwide higher education subsector. The factors that influence and contribute to global competition in higher education are complex and multifaceted. In addition, competition is manifested in many different forms—between universities within one country, between universities internationally, and between countries as a whole—and the various actors involved utilize a wide array of policies and strategies, official and unofficial, to increase their competitive advantage.

The chapters presented here indicate that embracing a spirit of competitiveness in higher education has its costs. The market-like forces that accompany the global competition phenomenon impinge on the higher education subsector in significant ways, resulting in serious concerns about their impact—in particular, their differential impact between and within countries and their HEIs. The recent focus on global university rankings has pitted institutions and countries against each other; all are playing a game that not everyone can win, and for which no clear rules have been established. Countries, institutions, and international student mobility are organized into core and periphery zones, based on historical factors (such as colonialism) that are exacerbated by neoliberal trends. Some of this organization is changing, and it must be emphasized that individuals, institutions, and nations do have agency; yet, the fact that some players start from a "lower" status initially must be taken into account when analyzing the impact of competition in higher education.

Moreover, the commercialization and commodification that accompany the global competition phenomenon represent serious risks to the missions of equity and access within HEIs worldwide. As many of the authors have pointed out in this volume, the trend toward competition raises serious social justice concerns—precisely because of the uneven playing field. If higher education is expanding, or the quality of institutions is improving, who is benefiting? Furthermore, even if quality can be "assured," the question remains: whose measures and standards are chosen to achieve and evaluate quality?

Competition may be an inevitable aspect of higher education reform in the future, but it need not be embraced without caution and foresight. Indeed, its negative impacts and risks must be investigated and addressed. The authors in this volume provide a wealth of insights from around the globe that serve as a starting point for meaningful, critical discussion about the tensions and challenges inherent in the global competition phenomenon.

# References

Harman, Grant, and Kay Harman. 2008. "Strategic Mergers of Strong Institutions to Enhance Competitive Advantage." *Higher Education Policy* 21 (1): 99–121.

Hazelkorn, Ellen. 2008. "Learning to Live with League Tables and Ranking: The Experience of Institutional Leaders." *Higher Education Policy* 21 (2): 193–215.

Huisman, Jeroen. 2008. "World Class Universities." *Higher Education Policy* 21 (1): 1–4.

Labi, Aisha. 2008. "Obsession with Rankings Goes Global." *Chronicle of Higher Education*, 55 (8): A27–A28.

Lingard, Bob. 2000. "It Is and It Isn't: Vernacular Globalization, Educational Policy, and Restructuring." In *Globalization and Education: Critical Perspectives*, ed. N. C. Burbles and C. A. Torres. New York: Routledge.

Marginson, Simon. 2006. "Dynamics of National and Global Competition in Higher Education." *Higher Education* 52 (3/4): 1–39.

Mohrman, Kathryn, Wanhua Ma, and David Baker. 2008. "The Research University in Transition: The Emerging Global Model." *Higher Education Policy* 21 (1): 5–27.

Slaughter, Sheila, and Gary Rhoades. 2004. *Academic Capitalism and the New Economy.* Baltimore, MD: Johns Hopkins University Press.

Usher, Alex, and Massimo Savino. 2006. *A World of Difference: A Global Survey of University League Tables.* Toronto: Education Policy Institute.

van Vught, Frans. 2008. "Mission Diversity and Reputation in Higher Education." *Higher Education Policy* 21 (2): 151–174.

# Chapter 2

# Higher Education Responds to Global Economic Dynamics

*Lynn Ilon*

## Introduction

As knowledge becomes the driving force for new wealth, higher education is viewed as a source of new knowledge creation, and it is rapidly evolving into a global economic industry. As such, higher education is moving away from policy-driven to economic-driven change. These changes are a result of global economic forces that were largely unleashed with the merging of first, second, and third world economies. This merge resulted in a new logic of production that rewarded corporations for their use of skilled knowledge while downplaying traditional advantages of location and access to raw materials. The global economy's further evolution into an emphasis on learning processes will foster additional changes in higher education.

In this chapter, my intention is to reset the sights of higher education planners and policy makers by outlining the global economic environment into which higher education has been thrust, and by discussing the specific components of economic change, which set the stage for higher education's move toward globalization. I begin with an overview of economic changes during the last century and how they began to impact the higher education subsector. Several changes in higher education resulting from the globalization process are then described and linked to the global economy. I end by outlining upcomi ng changes that are likely to further impact higher education.

# A Globalizing Economy

During the Cold War, much of the welfare of the citizens in the "first" and "second" worlds was a result of the political stability and influence of their respective economic systems. As the Soviet Union broke apart, the "second" world economy disintegrated. Gradually, trade was expanded from a prescribed set of trading partners in either "worlds" to a global world of agents willing to buy and sell. Relationships among countries were defined as much by their trade ties as by their political ties. Figure 2.1 shows the rapid expansion of international investment (foreign direct investment, or FDI) between 1980 and 1995. In 1980, only about 1 percent of the value of production was invested in countries outside a company's home country. Fifteen years later, that percentage had more than doubled.[1]

With increasing trade openness and technological advances, the race was on to locate both cheap labor and raw materials. Talented workers with knowledge became a competitive advantage, and the price of the best workers increased rapidly (Nonaka and Takeuchi 1995). This was not the case with unskilled labor, however, where the supply (the world's poorest citizens, moving away from agricultural subsistence) was large and growing, as more countries joined the rush toward global capitalism. Figure 2.1 shows how the percent of total value of products is attributable to market services of knowledge-intensive industries over a ten-year period. In 1995, such services accounted for about 22 percent of the value of the production

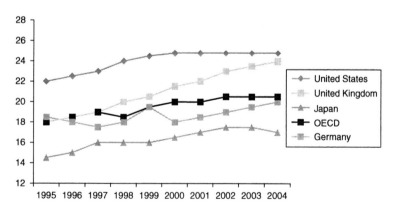

**Figure 2.1** Shares of Gross Value Added by Knowledge-Intensive Market Services, 1995–2004

*Source*: OECD (2007).

process. Ten years later, market services accounted for about 25 percent of services. The growth for other countries was even steeper. Knowledge-generated inputs such as market services account for a higher percentage of the total value of products.

And so, the revolution that had started with the downfall of the "second" world landed at the steps of higher education. Countries that wanted to move up the ladder of economic growth needed to shift from being the provider of unskilled labor, to the seller of the most highly skilled labor it could afford to educate. Skilled workers generate higher taxes; spur entrepreneurship; attract foreign direct investment; educate the next generation; and invest in social, cultural, and economic infrastructure (Porter 1998). Individuals who wanted to assure their livelihood needed to secure a higher quality education and to continue learning beyond formal schooling. Whereas higher education had generally been viewed as a service industry, providing a stable set of corporate and government leaders, it was now thrust into the world of highly competitive global production. Professors who were once hired as experts for the duration of their working life were being outpaced by the innovation of knowledge industries, and young digital natives took their learning elsewhere (Prensky 2005). Bill Gates, along with many other digital entrepreneurs, left higher education in order to become industry leaders. Higher education had a lot of catching up to do.

# International Competitiveness and Higher Education

These economic forces set the stage for higher education. Effectively, they are moving at least a part of higher education out of the government service sector and into the competitive sector. The line between public directing and corporate benefits has never been entirely clear in education—especially in higher education. But this line has been considerably blurred since higher education has become part of the competitive sector.

The West's notion of higher education has generally been that it serves a "higher calling" than commerce (Geiger 1994; Altbach 2007). Higher educational research was often thought of as having a broadly applicable benefit for society, while the application of such research often came from industry. But much of this seemingly clean division is not so clear in a globalized, knowledge-based world. For example, the simple ability to create knowledge, to learn rapidly, and to innovate has direct value for companies whether such abilities generate theories or products from universities. Industry is now often a producer of knowledge, or uses knowledge heavily. A country whose workforce is generally at a high level of skill and knowledge has a much higher standard of living for its people than does a society that is at the low end of a skilled labor force. Low skilled labor countries, or countries with substantial pockets of uneducated workers, carry a substantial burden, because their productivity as members of an advancing economy has become increasingly marginalized (Reich 1991). Thus, higher education, which explicitly aims to raise the overall productivity of its skilled workforce, is, in fact, an institution that assists industry in raising the overall standard of living.

Hence, one of the first policies in higher education—implemented in much of Asia—was to send the "best and the brightest" to the most prestigious schools the West had to offer, and later to bring them home to build their own higher education institutions. Keeping in mind that most Western countries already had a substantial number of foreign scholars in 1995, the Organisation of Economic Co-operation and Development (OECD, 2008) estimates that the number of foreign scholars has continued to increase about 3 percent per year during the subsequent ten-year period throughout surveyed countries. Korea and India have sustained a particularly high growth rate of scholars studying abroad—10 percent and 8 percent, respectively. This is a substantial investment in higher education, although, in the case of Korea, a fair number of its scholars stay abroad to undertake research.

These investments are paying off in big ways for some countries. Rodney Ramcharan (2004), Domingo Echevarria (2009) and P. E. Petrakis and D. Stamatakis (2002) all show that investment in higher education, particularly once a country has moved beyond the lowest levels of income and development, is a prime predictor of economic growth.

The Global Competitiveness Index (IIMD 2008) makes a similar distinction. This index rates countries on their overall competitiveness with categories of the poorest countries, middle-income countries and wealthy countries. In their view, the poorest countries are driven largely by their ability to produce raw materials and simple products—factors of production. The wealthiest countries are driven much more by innovation. In fact, the percentage of competitiveness attributable to innovation rises from 5 percent to 30 percent of overall competitiveness when one moves from a poor country to a wealthy one (World Economic Forum 2008, 7). To the extent that innovation is linked to higher education, then, higher education is not only a driver of innovation, but it is also a critical component of staying economically competitive within the world's wealthier countries.

The Global Competitiveness Index looks at factors related to competitiveness, such as ability to innovate and to absorb technology. Higher education is a primary factor in the ranking of competitiveness. Comparing India, Korea, and Singapore, ranking on higher education (one of eight sub scores), is higher than all other sub scores; Singapore and Korea get nearly perfect scores in this area, with India not far behind.

All this investment in higher education is paying off for some countries. An index designed to measure the knowledge productivity of national economies shows that eight of the top 30 countries are not from the West (World Bank 2008).

# Growth of Higher Education in Non-Western Countries

Although investment in highly qualified people who are educated abroad proved to be a good beginning strategy, that trend could not continue. As Asian (and Central European) economies began to move toward knowledge production, they needed to find a means of educating their own citizens at the higher levels, and demand for higher education increased. Accordingly, "the number of tertiary students worldwide doubled in size between 1975 and 1995 from 40.3 to 80.5 million" (SCUP 2008). UNESCO (2005) proposes that "the student population could reach 150 million in 2025" (84).

The *Times Higher Education* world ranking (*THE* 2008) indicates that nearly a third of the top universities in the world are now in the Asia Pacific region. This ranking reflects the vast investment in higher education undertaken by the region in recent years—and that investment continues. Entire countries see their future as chartered by higher education—at least in part. In the late 1990s, for instance, Singapore began an initiative to attract the presence of world-class universities into the country, and developed a kind of world-class hub of education. Much of the research being carried out in this hub of (largely Western) universities is linked to industry (Olds 2007). Korea has recently announced a "World Class University Project" designed to attract top researchers into their best universities (Materials Research Society 2008). In addition, the Korean government is offering subsidies to colleges that teach in English, and is supporting efforts to lure more foreign professors (McNeill 2008). Dubai is building a "Knowledge Village" to house its world-class university hub; originally launched in 2003, it brings together a dozen top world universities into a common site for education—allegedly the best in the Middle East (Zawya.com 2009; Mills 2008).

This is a sure sign that higher education is moving from a service to a competitive industry. Highly competitive universities in the West are taking their knowledge production on the road to populations who, heretofore, had to immigrate to the West to get such an education. Meanwhile, many countries are putting significant investments into such education hubs, or building their own internal ability to produce world-competitive education.

## Internationalization of Higher Education Students

Higher education students have often traveled the world for an education, but recently that trend is growing even faster. As an example, figure 2.2 shows the percentage of doctorates that have been awarded to non-U.S. students since 1997. In 1997, about 12 percent of all doctorates were awarded to foreign students. By 2007, this percentage had nearly tripled. More than a third of all doctorates awarded in the United States now go to non-U.S. students. It should be noted that this trend continued to rise even though, after the World Trade Center bombings in 2000, there was a one- to two-year downturn in the number of student visas issued by the United States. Thus, doctoral production of foreign students continued to rise.

Figure 2.3 gives us some idea of why this is the case. As mean incomes within a country rise (bottom axis), there is a concomitant change in the

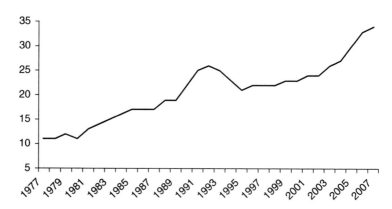

**Figure 2.2** Percentage of Doctorates Awarded to Temporary U.S. Residents, 1977–2007

*Source*: NSF (2008).

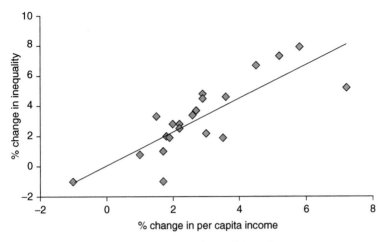

**Figure 2.3**   Relationship between Inequality and Growth

*Source*: Asian Development Bank (2007, 3).

inequality of incomes within the same country (left axis). That is, the rich get richer as the poor get (relatively) poorer. When low skilled labor is in large supply (many subsistence farmers have yet to trade their bare farm subsistence for formal sector labor, even as global trends suggest that many more will do so year after year), incomes at the bottom remain relatively low. But, as knowledge evolves and is reinvented each year, workers at the top of the knowledge scale must continuously renew their knowledge or fall behind. The number of places at the top of the knowledge tree for any given profession is thus limited. Short supply (with large profits to be made if a company has such workers) and high demand drive up the incomes of the world's elite knowledge workers.

Contrast this to higher education preparation a generation ago. In a world where economic opportunities were largely determined within a nation, and, perhaps, within a region of a nation, going to the local college and doing rather well was generally a ticket to a good income and a stable, steady job for life. In a global economy, only low-level and bureaucratic jobs are likely stable; the competition for top jobs is global and dynamic. Who would have guessed that health care could be globalized just ten years ago? After all, people access health care in their own communities. But there is a growing global health care industry in which the world's wealthiest citizens travel across the world to get special treatment and surgery at prices much cheaper than at their neighborhood hospital. Similarly, who would have thought that personal banking could be international?

The neighborhood bank took care of cash and loans twenty years ago. But nearly all banking is now national and, often, international. The work done in local branches is minimal, while the sorting, calculating, and managing go to Ireland or India.

The growing distance between economic haves and have-nots, along with a growing globalization of competition for the world's top jobs, means that the world's best and brightest go shopping internationally for higher education. Countries compete by assuring quality education for their residents. This used to mean that international students arrived in top Western universities; at this point, it means that universities throughout much of the world are competing for these students, and for world-class faculty.

## Internationalization of Research

The globalization of higher education and the increasing demand for education, especially at the top of the quality scale, is directly related to the internationalization of research. World dominance of top research no longer belongs exclusively to Western researchers. Competition to secure the world's best researchers is increasing—especially in the competitive fields of science and engineering, as figure 2.4 shows. This figure reveals that Asia is producing substantially more engineers than Europe, for example—nearly twice as many each year. It also has a substantial lead in natural

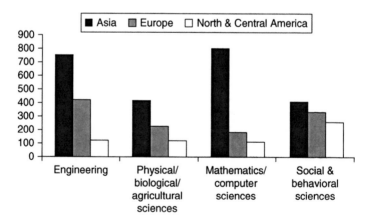

**Figure 2.4**  First University Science and Engineering Degrees by Region and Field, 2004 (in thousands)

*Source*: NSF (2008, 2–38).

sciences, and trails Europe only in mathematics majors. The pool of talent from which Asia can pull to train top-notch researchers in science and engineering is substantially larger than the pool for Europe. At one point, these Asian researchers would have stayed in Europe or North America to work, but today, they are returning to their home countries to good paying jobs and well-funded research settings.

Indeed, research is no longer the province of any one setting. The percentage of internationally co-authored articles continues to rise year-by-year. The only country where there has not been an increase is in China. One might speculate that China is now beginning to publish among its own researchers as much as it is within the international community of researchers (OECD 2007). The payoffs are large for high quality research no matter where it takes place. The best knowledge products (pharmaceuticals, new technologies, innovative software) can garner substantial profits from a world hungry to be at the top, compete, and stay healthy while doing so. The annual budget of many multinationals far outstrips the value of production of many countries. In a world driven by economics, the power of these large firms rivals the power of many smaller nations. Thus, the economic well-being of ever-increasing numbers of people are determined more by the economic context in which they work and live than by the political trends of their country or domicile. And, if the politics of their home country are not conducive to their lifestyle, they are part of a global class of workers who can move to another location.

## Finance and Investment in Higher Education

One of the ironies of higher education financing and investment is that, while Asia and the Middle East increase their funding of higher education from national sources, North America and Europe have faced continuing declines in investment for the last twenty years. Thus, the seeming financial crisis in higher education is defined more regionally than globally. Figure 2.5 shows the results of a survey done of multinational corporations by the OECD. They asked corporations which factors affected their location decision for research and development (R&D). The results are instructive for universities. The number of qualified R&D personnel, the presence of universities, and the ability to collaborate with universities were three of the top four reasons given for choosing an R&D site. University presence and influence was more important than tax breaks and legal issues, and even more important than growth potential. University presence is, by far, the most highly desired characteristic for location of R&D.

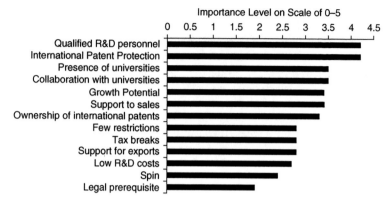

**Figure 2.5** Factors Affecting Research and Development Investment in Industrialized Economies

*Source*: OECD (2008, 69).

This kind of evidence would seem to argue for the investment in universities because their presence attracts the kinds of investments that put top money-earners to work. This generates tax revenues, increases demand for high-level services, and creates a region of economic well being. The economic rationale for higher education investment is clear. Yet, in much of North America and Europe, publicly funded institutions are suffering from declines in government spending, and, when they cannot garner replacement funds from private research investment, their quality is deteriorating (Johnstone and Aurora 1998).

As a global industry, higher education is expanding with partnerships between industry and government. In Asia and Central Europe, these private funds are being used to upgrade older systems and invest in new ones—an expansion of quality and quantity. As higher education merges with industry to form a competitive advantage to individuals, their communities, and their countries, the "local" meaning of higher education changes. At the upper quality of higher education, competition is not just global; it is aligned with industry and national investments. The underlying sustaining strategy focuses on the global but also includes strategic positioning of industries, analysis of world region competitiveness, and careful thought to which faculty and students it wants to attract. This same trend impacts colleges and universities at lower levels in a different way. Parents at the highest income levels no longer see their local college as an investment that works for their children. Nor are they necessarily invested in the future of their neighboring community. Collective support for tax dollars directed at local investment of education is giving way to the need to save for private investment in one's children's education—wherever that might take place. This reflects not just a loss of support for local education spending, but also represents a loss of identity with the local community, its future, and its residents.

## Toward Learning Environments

These trends impact higher education in fundamental ways. Failure to understand these changes in light of their underlying global economic roots leaves higher education planners and policies makers without the tools needed to adequately anticipate and plan for education in the future. Indeed, many still attempt policy solutions for issues that arise largely from global economic factors. Often, these issues are only loosely coupled to higher education policy—if at all. As a result, trends are noticed and tracked but cannot be predicted because the underlying economics remains hidden.

The revolution in knowledge economics is far from over. Indeed, it is likely just at its beginning stages. Higher education will have to make massive adjustments over the next decade in order to adapt to this changing environment, due to upcoming trends. First, technology is rapidly changing. Technology creates learning efficiencies for some skills and topics. But that change is modest relative to what is to come. For the first time in history, technology is changing the nature of how researchers can collaborate across time, distance, and space on joint projects. This allows for more efficient and timely research but also allows for a type of research that was heretofore unimaginable. People who are trained and whose life experiences are diverse can begin to share ideas and create solutions that are more robust than those coming from homogenous populations of researchers.

Technology is reminding us that learning is both a "soft" and a "collective" activity. That is, at some level, learning environments are not primarily based on hard infrastructure. Soft infrastructure—networks of people, management of information, access to data, and government policies that release information and foster innovation—has become dominant (Lundvall and Johnson 1994). An enabling environment of soft infrastructure might well mean that the university infrastructure of classrooms, textbooks, libraries, and even professors' offices and administrative units may take a distant second place, and be viewed primarily as support for soft infrastructure. Colleges are well advised to rethink the balance of hard and soft infrastructure investments in this new technological age.

Second, learning has potentially more economic value than does knowledge. Whereas knowledge is static—a matter of what is known at any given moment—learning is dynamic. Any given piece of software, for example, is only as valuable as it is to a market at the time of release and for a relatively short shelf life. Its real value is that its development has produced a team of people who know how to innovate together and build new ideas. Their ability to learn is the sustainable value of development. It is this learning potential that keeps a company in the knowledge business. Higher education will follow this new emphasis from knowledge to learning, and institutions will need to exchange their "expertise" model (in which the professor has knowledge, and students acquire this knowledge) for an active learning environment that is jointly managed by the experienced learner (the professor, perhaps) and the novice learner (the student). Shared learning strategies and development of new learning strategies will be the hallmark of a good education.

Third, the notion that colleges and universities exist in a relatively isolated world of public research will give way to a new role for universities: that of knowledge networker. Industry is already working closely with universities, and the difference between applied and theoretical research is blurring in the process. Does research that helps reduce the human cost of HIV and AIDS fall into the category of publicly beneficial research because it saves lives, or does it fall into the category of strategic research because countries and industry need to stabilize fragile parts of the world in order to reduce global costs and create markets? Each sector has a role to play in research: government, civil society, higher education, for-profit corporations and not-for-profit conglomerates. The university, reconceived as a central hub or network of people working together, means that higher education is not as much a physical place, as a virtual space of networked people.

Fourth, the higher education market will continue to globalize and get more competitive. Researchers, professors, and students will be recruited globally, and the best of them will see the market value of their work rise faster than typical wages in the economy. Top students may even be able to count on a free ride throughout their college years and, possibly, even find they exit with money in the bank. Given an environment that is dynamic and is increasingly defined by the ability to learn (not necessarily by a given body of knowledge), top students are no longer a cost to universities, they are a prime resource.

The lesson to be learned here is that much of higher education is shifting from a local service to a globally competitive industry. This shift is not a result of higher education policy makers and their decisions. There are specific, notable, and growing global economic forces that underlie this trend. Knowledge of economic forces empowers higher education planners to help their institutions adapt, and even thrive. There is a role for policy, but that role is changing. Just as all industries need oversight to assure that they operate within a purview that benefits the social, political, and economic goals of the society, so too does the higher education industry need regulation and oversight. How will learning access be provided by a government when higher education takes on its fullest "soft" infrastructure form? How can information be widely shared without discouraging its authors? How can diverse learners share knowledge across cultural divides? These questions will not be addressed in the policy debates of today, which still center on institutional growth. Rather, they will emerge as the new industry recognizes its global parameters and begins to build a global vision of knowledge, research, innovation, and learning. Each institution, then, will find its place within this global network.

# Note

1. Those numbers have since rapidly increased. By 2008, foreign direct investment had grown to 15 percent of all investment worldwide. Global FDI investment was valued at about 28 percent of the annual gross domestic product for all countries combined (UNCTD 2008).

# References

Altbach, Philip. 2007. "Peripheries and Centres: Research Universities in Developing Countries." *Higher Education Management and Policy* 19 (2): 106–130.

Asian Development Bank. 2007. *Inequality in Asia*. Manila: Asian Development Bank.

Echevarria, Domingo. 2009. "The Contribution of Higher Education to Economic Development in a Globalized Environment." PhD Diss. (forthcoming), Florida International University, Miami.

Geiger, Roger. 1994. *Knowledge and Money: Research Universities and the Paradox of the Marketplace*. Palo Alto: Stanford University Press.

International Institute for Management Development. 2008. *World Competitiveness Yearbook*. Lausanne: IMD. http://www.imd.ch.

Johnstone, D. Bruce, and Aurora Alka. 1998. "The Financing and Management of Higher Education: A Status Report on Worldwide Reforms." Paris: UNESCO World Conference on Higher Education. http://www.bc.edu.

Lundvall, Bengt-äke, and Björn Johnson. 1994. "The Learning Economy." *Industry and Innovation* 1 (2): 23–42.

Materials Research Society. 2008. *World Class University Project.* Warrendale: MRS. http://www.mrs.org.

McNeill, David. 2008. "Korea: The Government Is Offering Subsidies to Colleges That Teach." *Chronicle of Higher Education* 54 (28): A1.

Mills, Andrew. 2008. "Emirates Look to the West for Prestige." *Chronicle of Higher Education* 55 (5): A1.

National Science Foundation. 2008. "Science and Engineering Indicators 2008, Volume 1." Reston, VA: National Science Foundation. http://www.spaceref.com.

Nonaka, Ikujiro, and Takeuchi Hirotaka. 1995. *The Knowledge Creating Company.* Oxford: Oxford University Press.

Olds, Kris. 2007. "Global Assemblage: Singapore, Foreign Universities, and the Construction of a 'Global Education Hub'." *World Development* 35 (6): 959–975.

Organisation for Economic Co-operation and Development. 2005. *OECD Science, Technology and Industry Scoreboard 2007: Innovation and Performance in the Global Economy.* Paris: OECD.

———. 2007. *OECD Science, Technology and Industry Scoreboard 2007: Innovation and Performance in the Global Economy.* Paris: OECD.

———. 2008. *Recent Trends in the Internationalisation of R&D in the Enterprise Sector.* Paris: OECD.

Petrakis, P., and D. Stamatakis. 2002. "Growth and Educational Levels: A Comparative Analysis." *Economics of Education Review* 21 (5): 513–521.

Porter, Michael. 1998. *Competitive Advantage: Creating and Sustaining Superior Performance.* New York: Free Press.

Prensky, Marc. 2005. "Listen to the Natives." *Educational Leadership* 63 (4): 8–13.

Ramcharan, Rodney. 2004. *Higher or Basic Education? The Composition of Human Capital and Economic Development.* International Monetary Fund Staff Working Papers 15–2. Washington, DC: International Monetary Fund.

Reich, Robert. 1991. *The Work of Nations.* New York: Vintage Books.

Society for College and University Planning. 2008. "SCUP Trends in Higher Education." Ann Arbor: SCUP. http://www.scup.org.

*Times Higher Education.* 2008. "World University Rankings, 2008." London: *THE.* http://www.timeshighereducation.co.uk.

United Nations Conference on Trade and Development (UNCTD). 2008. *World Investment Report, 2008.* New York: United Nations.

UNESCO. 2005. "Towards Knowledge Societies." Paris: UNESCO. http://unesdoc.unesco.org.

World Bank. 2008. *Knowledge Economy Index 2008 Rankings.* Knowledge for Development Program. Washington, DC: International Bank for Reconstruction and Development.

World Economic Forum. 2008. *The Global Competitiveness Report, 2008–2009.* Geneva: World Economic Forum.

Zawya.com. 2009. "Dubai's Status as World-Class Education Hub Will Rise Further Once All Universities Shift to DIAC after 2010." Dubai: Zawya, January 14. http://zawya.com.

# Chapter 3

# Global Comparisons and the University Knowledge Economy

*Simon Marginson*

## Introduction

The first *Academic Ranking of World Universities* was issued by the Shanghai Jiao Tong University Institute of Higher Education in 2003 (SJTUIHE 2009), followed by the *Times Higher Education* rankings in 2004, which tried to put a British stamp on the emerging global order (*THE* 2008). Both of these triggered the transformation of world higher education.

In 2002, there were no global rankings. There were league tables and other comparisons of performance in some national systems, but little had developed globally. Comparative publication and citation metrics were of interest only to specialists. No one was talking about global classifications of institutions or cross-country comparisons of learning. Institutions were not globally referenced.

But things have since changed. Starting in 2003, global rankings began drawing media and public focus. World rankings began to affect the strategic behaviors of university leaders, governments, students, and employers (Hazelkorn 2008), leading to actions such as the accelerated investment in research and development (R&D) in China and Germany, and French mergers in higher education. Rankings have broken through all barriers, insinuating themselves in every university system (Sauder and Espeland 2009). The United States is an exception to this, given its fixation on the internal rankings of domestic universities and colleges in the *US News and World Report*, but its interest in global rankings will quicken if the East Asian nations, especially China, advance quickly in the top group of universities.

In one respect, the first worldwide rankings took the North American culture of comparison (present in *US News* and *McLean's Guide to Canadian Universities*) to the global level, with similar normalizing and homogenizing effects. But the global university ranking systems also offer a different kind of comparison. They have synchronized closely with worldwide developments in the knowledge economy, in the codification and valuation of knowledge, and with knowledge producers for all audiences. This synchrony is the topic of this chapter.

## The Global Knowledge Economy

The global stock of knowledge is the knowledge that enters common worldwide circuits, known as "global knowledge flows," and is subject to monetary and non monetary exchange. It is a mixture of both (a) tradeable knowledge-intensive products (from intellectual property [IP] and commercial know-how, to some industrial goods), and (b) knowledge goods freely produced and exchanged. Taken together, the production, exchange, and circulation of research, knowledge, and information constitute what can be called the global knowledge economy, or *k-economy*. The k-economy overlaps with the financial economy and industrial economy at many points. Yet, while knowledge eventually finds its way into innovations in production, and the economic role of knowledge-intensive goods is growing rapidly, commercial IP is only a small part of the k-economy, and it is not wholly contained by the old industrial descriptors. The k-economy is partly shaped by status competition which has always been integral to research and research universities. The k-economy also includes the semi-bounded, system-driven anarchy of open-source knowledge and cultural production, shading into social networking via the Internet and the continuously forming global civil society. To understand the k-economy, it is essential to grasp the extraordinary dynamism of open-source knowledge.

Knowledge goods have little mass, given their form as ideas and know-how and as first creations of works of art (that is, as original goods), and their production is ecologically sustainable. It requires little industrial energy, resting instead on human energy and time. Subsequently, most such goods can be digitally copied with minimal resources, energy, and time. Yet they can also be digitally reproduced as standard commodities for sale, acquiring prices and absorbing more energy. The production of commercial digital goods is subject to scarcity, while freely created digital goods are not. There is no natural scarcity of free knowledge goods, given that they multiply in dissemination. The condition of freely produced and circulated knowledge goods is hyper-abundance rather than scarcity; this is very different from conventional industrial production.

Thus, the k-economy is powered by two heterogeneous sources of growth. The first is economic commerce, which turns knowledge (along with everything else) to its own purposes, without exhausting the potentialities of knowledge. The second is free cultural creation: decentralized, creative, and unpredictable, freely circulating knowledge goods. Here, the production and dissemination of knowledge goods converges with the extension of communications and expansion of markets. According to Manuel Castells (2000) the unit benefits of networks grow at an increasing rate because of an expanding number of connections. Meanwhile, the cost of network expansion grows in linear terms. The cost of each addition to the network is constant. The benefit/cost ratio continually increases, so the rate of network expansion also increases over time until all potential nodes are included. Hence, the extraordinary growth dynamism of the open-source ecology, which expands much faster than the population or economic production, as does its quasi-democratic tendency to universality. In some countries, over 70 percent of households have personal computers; broadband access was at 25 percent in the Organisation for Economic Co-operation and Development (OECD) countries in 2006 and rising steeply, and blogs are growing exponentially (OECD 2008c, 55–62).

Meanwhile, the global roll-out of communications stimulates commerce. The grid of the network morphs into a product market and system of financial exchange, while open systems contribute a continuing flow of further knowledge goods originating from outside the trading economy. Some knowledge goods become captured by market producers and are turned into commodities. Others posit further acts of creation, communicative knowledge catalyzing new knowledge without mediation.

Is open-source knowledge emancipatory? The skills and hardware are not distributed universally, but the technological capacity to produce and exchange knowledge and cultural forms is globally in the hands of growing numbers of school children. It is much more widely distributed than the capacity to produce industrial goods or to invest at scale.

## Interpretations of Open-Source Knowledge

How can we understand open-source knowledge? Paul Samuelson (1954) systematized the notion of "public goods" as non-rivalrous and non-excludable economic goods that are under-produced in commercial markets. Goods are non-rivalrous when they can be consumed by any number of people without being depleted, such as knowledge of a mathematical theorem. Goods are non-excludable when the benefits cannot be confined to individual buyers, such as law and order. Paul Romer (1990) developed endogenous growth theory to explain the role of technological knowledge not just as saleable intellectual property but as a "non-rival, partly excludable good" (71) constituting conditions of production throughout the economy. Joseph Stiglitz (1999) argued that knowledge is close to a pure public good. Except for commercial property such as copyrights and patents, the natural price of knowledge is zero. Stiglitz also noted that a large component of knowledge consists of global public goods. The mathematical theorem is useful all over the world. Its price everywhere is zero.

But while economics has the tools to describe individual knowledge goods, it cannot yet fully comprehend a relational system (if indeed it is a "system") that lies partly inside and partly outside of cultural industries, publishing markets, and learned academies; in which exchange is often open-ended and populated by a strange public/private mixture of e-business and gift economy; and with information flows and networks tending to infinity. Notions of knowledge as a public good do not capture the scale, fertility, and disorder of the open-source regime. Samuelson (1954) saw public goods as pre-capitalist. They would be taken into markets and become private goods as technologies advanced. Yet open-source knowledge seems more post-capitalist.

In the Internet Age, the first move of economists was to model the quickly expanding stock of free knowledge goods simply as the source of commercial products. But not only does most knowledge never become a commodity, even knowledge goods in their commercial form are a peculiar beast, shaped by the logic of public goods. Knowledge goods are naturally excludable only at the moment of creation. The original producer holds first-mover advantage. This is the only solid basis for a commercial intellectual property regime. First-mover advantage diminishes and disappears once commercial knowledge goods are placed in circulation and become non-excludable. Any attempt to hold down commodity forms at this point is artificial. Copyright is not just difficult to police, it is violated at every turn and is impossible to enforce. Yet, free knowledge goods, which are so hard to alienate as property, are increasingly crucial as the basis of innovations and profitable new products in every other sector.

In 2008 the OECD swung the primary focus of policy on university research from commercial IP to open-source dissemination, noting that "a common criticism of commercialization is it takes at best a restricted view of the nature of innovation, and of the role of universities in innovation processes" (OECD 2008c, 120).

Intellectual property rights (IPRs) raise the cost of knowledge to users, while an important policy objective might be to lower the costs of knowledge use to industry. Open science, such as collaboration, informal contacts between academics and businesses, attending academic conferences and using scientific literature can also be used to transfer knowledge from the public sector to the private sector. There have been very few universities worldwide that have successfully been able to generate revenues from patents and commercializing inventions, partly because a very small proportion of research results are commercially patentable. In addition, pursuing commercial possibilities is only relevant for a select number of research fields, such as biomedical research and electronics (OECD 2008b).

The production and dissemination of "open science" leads to innovation. For the most part, commercial realization of the discovery process is better left to the market. Universities should do what they and only they are best at: curiosity-driven creativity and research training. Universal commodification will not happen. The highpoint of neoliberal expectations about academic capitalism has passed. Along with the neoliberal advocates of commercial university science, the critics of academic capitalism, too, will need to revise their position. We can thank the Internet and venture a little optimism here: the possibilities for freedom are opening. Not that we are on the brink of creative utopia: for every move to global extension and openness, there are a dozen attempts to secure closure. Global ranking is one such closure. Perhaps this openness/closure antinomy is integral to global agency.

# Regulation of Knowledge

Are all communications, cultural creations, and knowledge in the form of goods really equivalent in value? Economics says yes, given that all have zero price, but no one believes this. Goods without market economic value can be differentiated in ways other than price. In the open-source environment, knowledge flows both freely and disjunctively. When it passes through institutional settings and publications systems, it becomes structured and regularized, and channeled and restricted in flows that are often one-way. In the process, it acquires new social meanings. The means of knowledge production are concentrated in particular universities, cities, national systems, languages, corporations, and brands with a superior capacity in production or dissemination, mostly located in the United States and United Kingdom

(Marginson 2008). They stamp their presence on the k-economy and pull its flows in their favor. Knowledge is shaped and codified in research grant and patenting systems, research training, journals, books and websites, research centers and networks, professional organizations, and academic awards. These exercise a provisional authority in relation to open-source knowledge, without fully controlling it.

But *how* do the chaotic open-source flows of knowledge, with no evident tendency toward predictability, let alone equilibrium, become reconciled with a world of national hierarchies, economic markets, and institutions that routinely require stability and control in order to function? How is knowledge translated from the open-source setting into formal processes and institutions, so that these secure coherence and often a controlling role within the global k-economy? The answer is that in the k-economy, knowledge flows are regulated by a system of status production that assigns unequal values to parcels of knowledge and arranges them in ordered patterns. This system of status production has long-standing roots but has emerged in more systematic form in the wake of the Internet and the explosion in open-source knowledge.

The means of assigning status values to parcels of knowledge are league tables and other institutional and research rankings, publication and citation metrics, and journal hierarchies. Other ordinal rankings of outputs may emerge, such as comparative outcomes of student learning. For a long time, academic knowledge was structured by semi-formal procedures and conventions. Institutional ranks and journal hierarchies operated by elite consensus and osmosis rather than transparent and universal metrics. But in the last half decade, modernized, systematic, and accessible instruments have emerged primarily from the publishing industries, the Internet, and higher education itself—domains equipped to imagine global relations, though often with government support. These mechanisms appeared almost spontaneously and rapidly spread around the world, because they fulfill needs almost universally felt: to manage the formless and chaotic public knowledge goods, and to guide investments in innovation.

If the k-economy consisted solely of commercial markets in knowledge goods, then there would be no need to devise a status system for translating knowledge into ordered values. Market values expressed in prices would serve the purpose. However, most knowledge does not and cannot take a commercial form because of its public-good character. Markets cannot do the job, and other mechanisms for valuation are needed. In one sense, this is good for those who see a democratized communicative knowledge economy as a principal site for freedom. The heterogeneity of value serves to partly insulate the knowledge economy from the financial economy, and has the (not-incidental) effect of helping to sustain university autonomy and academic liberties. Notwithstanding the partial corporatization of higher education around the world, and the real loss of autonomy and freedom in some jurisdictions at the hand of both states and markets, the universities are outlasting neoliberalism. But every silver lining has a cloud. With commercial valuation unable to regulate most of the knowledge economy, we find ourselves using the other system of value already at hand: traditional university status.

On top of the almost limitless potentials of open-source knowledge and communicative human association—with its flexible combination of loose ties, free individuality, and vast common space for ideas and engagements—status competition via university rankings and journal hierarchies is imposing a traditional brand of closure in which existing institutional and imperial interests are sustained. Status closure is more complete than marketized closure. No doubt it provides stronger forms of control over the chaotic potentials of globalization in knowledge and learning. The k-economy is post-capitalist, but it is also pre-capitalist.

According to Fred Hirsch (1976) and Robert Frank (1985), status competition involves financial exchange in economic markets—for example, celebrity human capital in cinema. Other status competition takes place without money changing hands: in higher education in Germany, for instance, tuition is free but there is a scarcity of spots for students in elite institutions. The crucial point is that because status goods are goods of position within a finite hierarchy, there is an *absolute limit* to the number of goods of high value (Marginson 2004). As Hirsch (1976) puts it positional competition "is a zero-sum game. What winners win, losers lose" (52). "Saying that a high-ranked position in society is a thing of real value is exactly the same as saying a low-ranked position imposes real costs" (Frank 1985, 117). Positional goods/status goods confer advantages on some by denying them to others. In higher education, the zero-sum logic shapes the differentiation of consumption between elite and non-elite institutions, creating unequal opportunity, and of production, creating uneven quality. Hierarchy is both necessary to status competition and is continually fostered by it. This fashions it as an instrument of closure strong enough to impose itself on the fecund diversity of higher education and knowledge.

# The Mechanisms

There is one less credible indicator in the Shanghai rankings: Nobel Prizes. The rest are defensible: the number of leading researchers as measured by citations, publications in *Science* and *Nature*, publications in leading disciplinary journals, and these outputs on a individual faculty member basis. The data have been extended to rankings in five broad disciplinary fields, and more specialized disciplines in science, medicine, and economics.

The work done by Shanghai has had a profound impact on global comparison. It has established principles of measurement of real outputs rather than reputation, with transparent and accurate data collection, setting a benchmark for other measures and rankings, and hastening the evolution of the k-economy itself. The fit between performance, data, and ranking position is strong enough for nations, universities, and (potential) doctoral students to use the Shanghai rankings as a guide to strategic planning and investment decisions.

With that said, there are problems in the use of the rankings. They are often read as a "world's best university list," not a research performance list. This is inevitable, especially in the absence of a credible all-round measure of performance that includes the quality of teaching and/or learning (Dill and Soo 2005). There are also widespread concerns about the natural bias of the measures in favor of English language nations, big science, and medical universities.

The Shanghai rankings have brought into the mainstream bibliometric data on research and citations, including impact measures and judgments about the centrality and quality of field-specific journals. The field of data compilation involves two major publishing houses and researchers in many nations specializing in science indicators. For example, in 2007, Leiden University in the Netherlands announced a new system based on its own bibliometric indicators. Its primary indicator is average impact measured by citations per publication normalized for academic field—in other words, controlled for different rates of citation in disciplines, and modified to incorporate institutional size, known as the "Crown Indicator" (CWTS 2007). Arguably, this provides the best comparative data on research performance so far, though all such metrics tend to block recognition of innovations in field definition and work in new journals.

The publication of the OECD's Programme for International Student Assessment (PISA) comparisons at the K-12 level triggered the possibility of something similar in higher education, and in 2008, the OECD Assessment of Higher Education Learning Outcomes (AHELO) project commenced (OECD 2008a). The potential impact of comparative objective measures of learning outcomes can hardly be overstated. At present the main means of measuring comparative learning outcomes, likely to receive further development (CHED 2008), are surveys of students or graduates but their subjective character limits these data. The AHELO project is piloting measures of the generic skills of graduates, graduate competence in two disciplines (engineering and economics), graduate employment outcomes, and contextual data to assist data interpretation. It is envisioned that the units of comparison will be institutions rather than national systems. Though the technical and policy obstacles to this project are formidable, there is much policy momentum in favor.

## Diversity and Closure

There is no one single "Idea of a University." There are many different missions, structures, and organizational cultures, all nested in national contexts, historical identities, and conditions of possibility. In the United Kingdom, Australia, and New Zealand, national systems combine university autonomy with explicit central steering. The Nordic/Scandinavian university combines high participation, social equity, research culture and university autonomy with strong state investment (Valimaa 2004, 2005). The German-style university opts for elite participation, research culture, and state administration. The Latin American public university fosters high participation, scholarly culture, and building the nation-state. The emerging science universities of China, Taiwan, Korea, and Singapore are produced by state investment and designed to secure global competitiveness. India at its best fosters high quality technology and business-focused institutions. Beyond the research university are strong vocational sectors in Finland, Germany (*Fachhochschulen*), France, and other nations. Across the globe, there are online institutions and specialized institutions in single fields.

Yet science universities and scientific knowledge in the Anglo-American tradition tower above everything else. If this partly derives from sheer knowledge power itself, it is a victory that has also been "earned" by excluding ideas and works of other traditions, taking the form of a self-reproducing hegemony. The first generation of rankings and journal classifications have seemingly sealed that process. They elevate not just Anglo-American knowledge but the institutional missions, habits, and assumptions of the leading Anglo-American universities.

The implications of normalizing ranking systems for the actual existing diversity in higher education, not to mention the potentials for future diversity, are a principal concern. Early criticism has been mounted more against bias effects than reshaping effects. One example of the effects of rankings and its research counts is the fate of the national public universities in Latin America. Both the Universidad Nacional Autonoma de Mexico (UNAM) and Universidad de Buenos Aires en Argentina (UBA) provide access to a quarter of a million students or more on many sites and perform many functions in national and regional development, and social and cultural life, as well as national research leadership. This range prevents UNAM and UBA from concentrating resources so as to maximize research intensity and reputation like Princeton or Caltech; in addition, scholarship in Spanish is unrecognized in the rankings. These institutions appear in the 151–200 bracket of the Shanghai rankings, yet these are great universities, and their long-standing model has distinctive strengths.

The reputational element in the *THE*, and its knack of changing the measures to elevate selected institutions, allows it to include the Indian Institutes of Technology (IITs) and push the Dutch technical research universities above traditional comparators, while more specialized and nationally bound vocational sectors such as the German *Fachhochschulen* remain excluded.

Given that front-rank scientific publication is almost exclusively in English, the knowledge status system is an English language system. However, open-source knowledge permits and produces greater plurality, and English is no longer the primary language of the Internet.

The use of institutional classifications as part of a system of comparison opens the way to plural comparisons in place of a single global ranking regardless of mission. Institutions of like mission and/or activity profile can be compared with each other. Research-intensive universities, specialist technical vocational institutions, and stand-alone business schools can be grouped with their fellows. This would enable more precise, less homogenizing comparisons and better identify the worldwide distribution of capacity in the k-economy. It would allow for the expression of several different hierarchies rather than one universal hierarchy of institutions (though most likely the research-intensive university hierarchy will tend to dominate the others). As yet, there is no global classification, but some of the building blocks are being put in place. The United States has Carnegie. China now has national classifications, and a classification is being developed for the 4,000 higher education institutions in Europe (van der Wende 2008).

## Conclusion

Research universities are subject to two systems for regulating value—values which operate alongside each other, and intersect only some of the time: the *economic value* of commercial intellectual property and knowledge-intensive products; and the *status value* of public good knowledge, determined by university rankings, research, and publication metrics (and probably by learning outcomes in the future). Research universities are pulled by both kinds of regulation, but their eggs are mostly in the second basket. A small portion of knowledge generates surplus revenues, and all knowledge in universities generates knowledge status-power within the k-economy.

Yet, the status of knowledge is calculated only for knowledge codified in refereed papers, monographs, and other formal mechanisms. A third category of knowledge remains outside the regulation of value: research, scholarship, ideas and other knowledge and information created in universities (and elsewhere) which is neither sold in a market nor counted and ranked in a status system—that is, works in the form of open-source knowledge. This knowledge always has the potential to feed into and condition the formal academic domain, while some feeds straight into the commercial domain, through consulting, "quick and dirty" research, and employee creativity, without being valued in academic terms.

Thus, the research university is pulled three ways: by the commercial imperative, by the formal knowledge status system (dominant within the university), and by the unpredictable swirls of open-source knowledge. These three heterogeneous "systems" are in highly unstable symbiosis, and more unpredictable changes will surely occur, with differing implications for national organizations, institutional forms, academic behaviors, relations of power, and the vectoring of the life-world.

Much critical analysis is focused on ongoing tensions between commercial and academic values in research (e.g., Bok 2003). Still, commercial research, while economically significant, constitutes a relatively small part of total research revenues and time. The more important tension is between open-source knowledge production, and the status hierarchy in knowledge and knowledge production fostered by rankings and metrics. Here the strong universities are building barriers to creativity, democratic communicability, and all-round global development.

"Competition" in the global higher education subsector is only loosely related to a neoliberal understanding of economic competition. "Comparison" is more pervasive than competition, and cooperation is also important. While the rise of the "global competition state" as the policy paradigm for national governments is one factor sustaining ranking systems, much more is happening than neoliberal ideological capture. The culture of comparison sweeping worldwide higher education originates primarily in global civil society rather than states, and is sustained by a global visibility that became possible only in the Internet era. These phenomena are new to the history of societies and organizations. The same comment can be made about communicative globalization, and that fast-growing and peculiar beast we euphemistically title the "knowledge economy," which is both more and less than a conventional economy.

University rankings do not constitute the ultimate horizon of possibility. They secure closure, but in the long run, any closure is unstable, especially enclosure of the fast changing global knowledge economy. Every global-system architecture is provisional and open-source possibilities are always bubbling up from below. Status is not the only game in town.

For systems and institutions, there is a raft of immediate issues of power and positioning. The manner in which global rankings are configured is a hot topic. Rankings and the associated regulatory processes have created a new policy space for global discussion in which civil organizations like the Education Policy Institute mix with universities involved in the science of ranking and many state and semi-autonomous agencies, especially in Europe and East Asia. Both OECD and UNESCO have been active in the early rankings conferences. So far these discussions have yet to devise a multilateral solution to the unilateral global effects of rankings that elevate United States and, to a lesser extent, UK higher education high above the rest.

Rankings will always elevate and reproduce the power of those who are already strong; however, ranking systems could be reworked to incorporate a greater plurality of language, institutional type, and mission. More use of a multiplicity of single indicator-based comparisons, in place of composite indicators and league tables based on them, would help. The more space made available for heterogeneity of valuation, the better. The ultimate answer lies in the domain of open-source knowledge outside rankings. The more creativity is sustained and communicated outside orthodox academic research and publishing, the greater the potential for "flat" and plural relations, and the more the knowledge economy starts to morph into its potential successors, which are the creative economy, and the society of ideas and design (Peters et al. 2009).

# References

Bok, Derek. 2003. *Universities in the Marketplace: The Commercialization of Higher Education*. Princeton, NJ: Princeton University Press.

Castells, Manuel. 2000. *The Rise of the Network Society*. 2nd ed. Oxford: Blackwell.

Center for Higher Education Development. 2008. *Study and Research in Germany*. Bonn: DAAD. http://www.daad.de.

Center for Science and Technology Studies (CWTS). 2007. *The Leiden Ranking*. Leiden: Leiden University. http://www.cwts.nl/cwts/LeidenRankingWebSite.html.

Dill, David D., and Maarja Soo. 2005. "Academic Quality, League Tables, and Public Policy: A Cross-National Analysis of University Rankings." *Higher Education* 49 (4): 495–533.

Frank, Robert. 1985. *Choosing the Right Pond: Human Behaviour and the Quest for Status.* New York: Oxford University Press.

Hazelkorn, Ellen. 2008. "Learning to Live with League Tables and Ranking: The Experience of Institutional Leaders." *Higher Education Policy* 21 (2): 193–215.

Hirsch, Fred. 1976. *Social Limits to Growth.* Cambridge, MA: Harvard University Press.

Marginson, Simon. 2004. "Competition and Markets in Higher Education: A 'Glonacal' Analysis." *Policy Futures in Education* 2 (2): 175–245.

———. 2008. "Global Field and Global Imagining: Bourdieu and Relations of Power in Worldwide Higher Education." *British Journal of Educational Sociology* 29 (3): 303–316.

Organization for Economic Co-operation and Development (OECD). 2008a. "Roadmap for the OECD Assessment of Higher Education Learning Outcomes (AHELO) Feasibility Study." IMHE Governing Board, Document Number JT03248577. Paris: OECD.

———. 2008b. *Tertiary Education for the Knowledge Society: OECD Thematic Review of Tertiary Education.* Paris: OECD.

———. 2008c. *Trends Shaping Education: 2008 Edition.* Paris: OECD.

Peters, Michael A., Simon Marginson, and Peter Murphy. 2009. *Creativity and the Global Knowledge Economy.* New York: Peter Lang.

Romer, Paul. 1990. "Endogenous Technological Change." *Journal of Political Economy* 98 (5): 71–102.

Samuelson, Paul. 1954. "The Pure Theory of Public Expenditure." *Review of Economics and Statistics* 36 (4): 387–389.

Sauder, Michael, and Wendy Nelson Espeland. 2009. "The Discipline of Rankings: Tight Coupling and Organizational Change." *American Sociological Review* 74 (1): 63–82.

Shanghai Jiao Tong University Institute of Higher Education. 2008. *Academic Ranking of World Universities.* Shanghai: SJTUIHE. http://ed.sjtu.edu.cn.

Stiglitz, Joseph. 1999. "Knowledge as a Global Public Good." In *Global Public Goods: International Cooperation in the 21st Century,* ed. I. Kaul, I. Grunberg, and M. Stern. New York: Oxford University Press.

*Times Higher Education.* 2008. *World University Rankings.* London: *THE.* http://www.topuniversities.com/.

Valimaa, Jussi. 2004. "Nationalisation, Localization and Globalization in Finnish Higher Education." *Higher Education* 48 (4): 27–54.

———. 2005. "Globalization in the Concept of Nordic Higher Education." In *Globalization and Higher Education,* ed. A. Arimoto, F. Huang, and K. Yokoyama. International Publications Series 9, Research Institute for Higher Education. Hiroshima: Hiroshima University.

Van der Wende, Marijk C. 2008. "Rankings and Classifications in Higher Education: A European Perspective." In *Higher Education: Handbook of Theory and Research,* ed. J. Smart. Dordrecht, The Netherlands: Springer.

# Chapter 4

# Accounting for Excellence: Transforming Universities into Organizational Actors

*Francisco O. Ramirez*

## Introduction

Throughout the world universities are increasingly engaged in activities that commit them to pursue excellence and account for progress toward excellence. Much of this accounting involves formalizing faculty assessments and standardizing university assessments. These assessments emerged in the United States earlier than Western Europe, but they are now worldwide. A crucial dynamic that facilitates this development is the transformation of universities from historically grounded and nationally specific institutions to organizational actors influenced by universalistic rationalizing models. As organizational actors, universities are expected to have goal and plans to attain them, as well as mechanisms for evaluating their progress. Universities are expected to act as if they can learn from other universities and from expertise on how to improve.

This chapter seeks to interpret the worldwide transformation of universities with respect to accounting for excellence. The first part of this chapter reflects on the rationalized university as an organizational ideal and its implications for "accounting for excellence" practices, particularly the rationalization of the university motivated by the neo-institutional perspective (Ramirez 2006; Frank and Meyer 2007; Meyer et al. 2007).

Next, the chapter focuses on faculty assessments by looking at a case study of the annual faculty report. I argue that this and related practices facilitate U.S. university participation in national and international rankings.

In the last section, I argue that U.S. universities underwent earlier organizational rationalization and differentiation in part because they were less differentiated from other social institutions. Absent the buffering authority of the state and the professoriate, U.S. universities became organizational actors dealing with multiple stakeholders in search of resources and legitimacy earlier on than their counterparts elsewhere. These accounting exercises have now surfaced and diffused within Western European universities, but they encounter resistance. The "need" to create more competitive universities is a feature of higher education discourse throughout the world. This "need" facilitates university participation in international rankings that then further enhance this "need." This chapter concludes with reflections on why the current globalization differs from earlier educational "borrowing" practices and is more difficult to resist.

# The Rationalized University

From its medieval roots the university evolved in the eighteenth and nineteenth centuries to become a national institution. For as long as national historical legacies held sway, the national university could invoke and embellish its unique national character to buffer itself from invidious international comparisons (see Flexner 1930 for an early cross-national assessment of universities). International comparisons presuppose standards, comparability, and the portability of "best practices," determined by professional experts.

An earlier phase of this development emphasized access to schooling, and national report cards were mostly about enrollment levels and enrollment growth. A more recent phase stresses achievement across a wide range of subject matter, from mathematics and science, to reading and civics. International tests like the Trends in International Math and Science Study (TIMSS) and the Programme for International Student Assessment (PISA) have proliferated and the number of countries undertaking these tests has also increased. National school reforms are replete with references to how well a nation's students did compared to other countries. The complex ways in which these tests are produced are discussed within academia, but policy discussions rarely consider methodological controversies, because educational reforms are usually grounded in faith in international assessments.

Not surprisingly, the same shift in focus is taking place regarding higher education, a shift from access and enrollments to quality and achievement. Inquiring minds seek to know what the contemporary monks and their acolytes are really learning. How does this learning contribute to the betterment of society? Who is doing it right? How can we learn from winners? If these questions seem farfetched, consider the following e-mail query from an international consulting firm to the author of this chapter:

> I am part of a team supporting the Kingdom of Bahrain on a comprehensive education reform (spanning all levels: primary, secondary, tertiary, and vocational)....I would be most obliged if we could spend a bit of time discussing your perspective on international systems of higher education. Specifically, I would appreciate your insights on:
>
> - What major trends are occurring in education internationally, particularly related to university institutions?
> - What are the factors that make an educational system most successful, particularly in a small, developing nation?
> - What nations are achieving the most success in their educational reforms? From whom can Bahrain learn? (Personal communication)

From a realist perspective all of this makes good sense. Why should a country or a university not learn from political and educational winners? Why should those seeking to upgrade themselves not figure out the efficacious technology others have utilized to thrive? The problem, of course, is that the relationship between means and ends is more complex and uncertain when looking at higher education and national development goals than when examining mousetraps and the demise or continued prevalence of mice. Dead mice unobtrusively testify to the efficiency of a mousetrap. But what exactly does one point to in showing the relative superiority of a higher education system? The most obvious and the most widely imagined national outcome of higher educational development is economic growth. But there is little systematic evidence that expanded systems of higher education promote economic growth (Chabbott and Ramirez 2000).

From a neo-institutional perspective, it is precisely this lack of evidence regarding efficaciousness that subjects national educational systems and universities to the influence of rationalizing external models to "get it right" and to success stories that presumably illustrate the models' main assumptions. What these rationalizing models achieve is greater isomorphism, though not necessarily greater efficacy. Through mimetic and normative dynamics, universities begin to resemble each other, especially those dimensions that reformers identify as organizational in character. Each university can continue to assert some symbolic distinctiveness as in sagas connected to its founding moments (Clark 1972), even as its curricula and faculty cease to be distinctive. In earlier eras the mimetic and normative dynamics were mainly contained within the national domain. University rankings within the United States, for instance, predate the international Shanghai and *Times Higher Education* (*THE*) rankings by decades. But the permeability of national boundaries, the decline of national state charisma, and the sense that states and organizations are embedded within a world society trigger international comparisons based on transnational standards.

The rationalization of the university is influenced by its transformation into an "organizational actor" and this transformation in turn facilitates further rationalization. Georg Krücken and Frank Meier (2006) use the term "organizational actor" to refer to "an integrated goal-oriented entity that is deliberately choosing its own actions and thus can be held responsible for what it does" (241). It is perhaps commonplace to imagine universities as organizations and indeed higher educational management is emerging as an area of research and practice. But universities qua national institutions often lacked organizational backbone or formal administrative structure (Musselin 1999). Older universities were instituted as communities with tradition (and its frequent reinventions) as a guiding light. Newer universities were established to look more like deliberate associations, but these were insufficiently rationalized as organizational actors.

Today, however, universities are expected to function as organizations, to have goals, and to have plans for attaining them. Universities are expected to have specialized personnel and smart systems to bring these plans to fruition. And lastly, universities are expected to collect and analyze data to determine how well they are performing. Performance assessments in turn led to refining goals, targets, resource allocation decisions, and strategies for more effectively attaining these goals. These circular processes enhance the sense that universities are organizational actors; acting like a rational actor has become the bottom line (Ramirez 2006).

## The Rationalization of the University

Universities as national institutions are more difficult to rank than schools, because each university can claim unique legacies or distinctive styles. The age of the university, of course, increases the likelihood that it can evoke an image of itself as a core feature of a national tradition. The more the national tradition is insulated from transnational expertise, the more the university persists as a national institution. In the twenty-first century, universities as distinctive national institutions are increasingly "at risk" of veering toward a model-driven formal organization. Diffuse goals such as broad accessibility and social usefulness are on the rise in national educational agendas. David J. Frank and Jay Gabler (2006) convincingly demonstrate patterns of curricular change and isomorphism that are attuned to the imperatives of broad accessibility and social usefulness. A third driving idea is that the university should become more organizationally flexible and effective. This idea often goes hand-in-hand with the notion that universities should be "free" to seek multiple sources of funding. This in turn blurs the boundaries between universities and a range of organizations and associations in civil society and industry.

The university qua firm has lead to a critical discourse on academic capitalism (Slaughter and Leslie 1997) or managerialism (Gumport 2000). This literature supports the premise that the global market or global capitalism is the visible hand that rocks universities as national institutions. The search for new revenue streams, for example, has lead to new programs to attract students from outside national borders. University-industry ties are frequently discussed in these terms. However, many changes in universities make little sense if profit were the bottom line. Cost efficiency does not drive efforts to create a more diverse student and faculty body, efforts well underway in the United States and elsewhere. Neither the relative triumph of the social sciences in university curricula (Frank and Gabler 2006), nor the global diffusion of women's studies in higher education (Wotipka and Ramirez 2008) are by-products of the global economy. Instead, the global cultural emphasis on accessibility invites diversity just as the social usefulness theme undergirds the rise of the social sciences. The university as an upper-class institution or as an institution for men is de-legitimated on universalistic cultural grounds. In varying degrees, universities have changed to appear to be more inclusive, more student centric, more socially useful, and more organizationally flexible and effective. To monitor progress along these different fronts, universities collect relevant statistical data, create specialized offices that both monitor and signal commitment to progress, expand curricular offerings, advertise the relevance of the university and its services, and engage in both faculty and university assessments. The latter increasingly involve international benchmarks by situating universities in the hierarchies or rankings generated via the Shanghai or *THE* university rankings.

Universities have frequently promoted and protected themselves, but only recently have these activities been explicitly framed in terms of transnational standards and international rankings (Engwall 2008). Much of the literature on how universities and governments utilize these rankings is limited to the first world. It is assumed that universities in less developed countries are unlikely to be "competitive," and the resource gaps between universities in rich and poor countries make rankings less relevant. But this turns out to be untrue. Universities in poor countries use the rankings to communicate symbolically high aspirations. Furthermore, the rankings allow them to make not only global, but also regional, and even national, comparisons. Depending on the reference group of countries, the rankings can be used to promote an image of the university as a high quality establishment or as one striving to attain high international standards. This organizational presentation of self can be aimed at governments, at other universities, at potential funding sources, and at alumni.

The scope of the rationalized university is truly worldwide in character. Krücken and Meier (2006) report that the website of the University of Botswana is loaded with the familiar descriptors of the rationalized university: centers of excellence, international orientation, quality management, lifelong learning, public accountability, and interdisciplinarianism. Resource-limited universities are often eager to link to world standards and international comparisons that legitimate themselves as organizational actors. Just as resource-limited countries often rely on external sources of legitimating, such as membership in the United Nations, resource-limited universities may be inclined to invoke their external ranking to validate themselves as "real" universities. Resistance may be more evident in older and more resource-rich universities but these too undergo organizational changes that align them with current visions of the effective university (e.g., see Soares 1999, who depicts organizational changes in Oxford).

# The Intensification of Rationalization: A Case Study

Universities in the United States are often held up as exemplars of universities that are broadly accessible, socially useful, and organizationally flexible. Their success in international rankings and in bibliometric contests is often attributed to their superior organization and management. Higher education reforms in other countries, directly or indirectly, are influenced by U.S. success stories and the principles and assumptions these stories illustrate. Many of these stories involve distortions, underestimating the role of federal and state governments in fostering and maintaining universities, and overestimating the differences between public and private universities. Most importantly, the causal tie between better organization and management and university success is usually based on anecdotal evidence.

But it is true that some forms of rationalization and accounting for excellence took place earlier in U.S. universities. Professors in the United States were subjected to student evaluations long before this practice took root in Europe. Earlier challenges about the competence of students to assess faculty quality are now dead. The U.S. professor is not expected to profess but to teach. The students do not assess the professors' mastery of theory or research methods but whether the course material was well organized and effectively communicated by a professor that truly engaged them. The actual content of the evaluation protocol varies, but the commonalities clearly outweigh the differences. Though elite universities clearly value scholarly achievements more than teaching assessments, teaching evaluations are routinely utilized in both elite and non-elite universities. Furthermore, as professors are recruited from one university to another they are often asked to display evidence of teaching effectiveness. The portability of ratings is grounded in the belief that good teaching is good teaching, regardless of the local context.

Of course, there are many other factors that go into the overall assessment of professors. I examine here one dimension of faculty assessment, the annual report, using materials from a highly ranked professional school in a highly ranked university. I highlight both what is expected and how these expectations have become formalized and elaborated.

The annual report is a common practice across universities in the United States. It is produced by individual professors and sent to school deans, department chairs, or departmental colleagues. In some universities, these reports are cited as the basis for determining the magnitude of salaries, but they are undertaken even in universities where salary raises are standardized, such as the California State University system. Even in universities where the magnitude of salary raises fluctuates, the degree of fluctuation within a year is modestly related to variation in the quality of professorial profiles as revealed in these annual reports. In universities where big annual raises are possible, academic market processes mostly trigger these raises. External offers from prestigious competitors are by far better leverage than attractive annual reports. External offers, of course, are influenced by judgments of quality based on performance. But if one is not movable for spousal or other reasons, one is likely to face more modest salary bumps.

Before we delve into the contents of an annual report, let us briefly reflect on why they take place. Since all faculty, including those with tenure, issue these reports, they are not instruments designed to weed out the weak or unproductive. Since the reports are not strongly linked to a system of rewards, monetary incentives are not the core drivers for report writing. It is misleading to think that universities seek this information to rationally reward the more productive.

The annual report, I contend, is first and foremost a symbolic affirmation of the university as an organization. More concretely, the annual reports activate the idea that professors are accountable to those who play managerial or administrative roles within the university, and furthermore, that accountability displays are orderly, standardized, and universalistic. Within the appropriate academic unit, every professor faces the same annual report criteria, and it generates data that creates the impression that the university is a data-driven organization in assessing faculty. However inconsequential these performance assessments may be, they give the idea that the organization is engaged in a fair and objective assessment of its personnel. The alternative to the rationalized organization is imagined to reek with subjectivity, arbitrariness, favoritism, and other organizational shortcomings. Thus, while the innovation called performance assessments produces "fear and trembling" in some European universities, annual reports do not generate the same adverse reaction among U.S. professors. The practice is so common and routine that it rarely evokes criticism or commentary.

What goes into an annual report? Not surprisingly it covers familiar ground: scholarship, research, teaching, and service. In earlier eras, a loosely structured narrative sufficed as an account of scholarship and research activities. The standardized teaching evaluations, however, have long been a staple in them. Students evaluate course organization, communication effectiveness, and engagement with students. For each of these dimensions, an average assessment score can be computed that can then be compared to the department or school average. Overall scores can also be computed and compared. Qualitative comments are solicited from students and this feedback presumably helps professors upgrade their teaching. Student anonymity is taken for granted.

The scope of annual reports has expanded, providing professors opportunities to demonstrate virtue across more varied dimensions. In this case study, seven types of data are solicited: publications, courses, committees, funding, service to the school and to the university, service outside the university, professional activities, and honors and awards. The narrative on scholarship is boiled down to a specification of research products. The underlying message is simple: a good scholar generates research products, differentiated into books, refereed articles, and chapters. Earlier versions of this format did not call for further specifications, but now one is asked to specify whether the product is single- or multiple-authored. In the case of multiple-authored papers, one is expected to indicate the order of authorship. Both of these more recent developments facilitate partitioning the credit due to the authors. These developments may seem out of line with an educational climate that celebrates collaborative endeavors, but they make sense as indicators of precise accounts of excellence. Though not as yet set forth as expected data, knowledge of the relative status of the collaborators is no doubt factored into assessments of scholarship. One's relative standing in the authorship line may be weighted differently depending on one's relative standing in the discipline or profession.

A related issue has to do with products that have yet to materialize. A sharp line is drawn between work in progress and work that has already been accepted but has yet to materialize. Both are recorded but only the latter count. The closer the accepted product is to publication, the greater the likelihood that pagination will already be known. Thus, professors are encouraged to cite pages when known. This too, is a recent further specification and one that gauges proximity of materialization with the more proximate, perhaps, counting more. (Papers under review count because they have materialized and can be viewed as product candidates.)

The sections on teaching and student committees include more than the standard student evaluation of teaching, reminding faculty to include data on the kinds and number of courses taught, on the different kinds of committees served, and on the different roles played within these committees. Professors are expected to have their own, presumably more accurate, records regarding the relevant materials sought. Faculty must explain why they taught fewer courses than the norm and why some course evaluations are not available. None of this is draconian, as there are well-established grounds to justify a low course load or not having a course evaluated. For the latter, new courses and courses with few students are exempt. For the former, administrative responsibilities or grant supported "time buy outs" are acceptable accounts. It is not clear what sanctions would be applied to those who deviate from these norms. What is clear is that the annual report activates norms regarding the value of teaching and does so in greater detail than a simple "we care about teaching" statement conveys.

The annual report activates norms regarding the quest for funding. In this domain, however, both submitted and received grants count. The former may be grant proposals that are pending or even grant proposals that have been rejected. It may seem odd that the latter count whereas works in progress do not. Perhaps a failed research proposal is a more concrete product than a work in progress. Or, perhaps funding proposals are more aligned with a collective good, or the support of doctoral students, than work in progress. Even a failed research proposal communicates that one at least tried!

The last sections of the annual report focus on service, professional activities, and honors and awards. The service category distinguishes between service to school, university, and the wider society. In a highly rationalized university there are all sorts of committees and, hence, all sorts of opportunities to display service. There are university and school-wide standing and ad hoc committees. Committee membership may be brought about through an election or an appointment or even via the old fashioned volunteer route. Beyond the university, service ranges from local, to state, to national, and even to international organizations/associations. Membership in a board of the National Academy of Sciences or in a Ford Foundation committee, or in some international counterpart counts, as do multiple other venues, where service can be undertaken and recorded.

It is equally easy to identify an array of professional activities, from presentations of conference papers, to reviewing journal papers, to serving on editorial boards. Some of these activities could be classified as service, such as evaluating research proposals for a foundation. Further rationalization down the road may facilitate more precise classification of one's activities. The simple rule is not to double count the service or professional activity. Lastly, the section on honors and awards has a dual purpose: it explicitly provides faculty with an opportunity to shine, and it allows the school to garner information that makes it shine. This becomes clearer when we turn to the question of university assessment and rankings.

## Ranking Universities: A Case Study

Universalistic standards co-vary with a world of international comparisons. Still, not all phenomena in this world are subjected to comparisons and rankings. We do not engage in ranking exercises as applied to races, religions, civilizations, etc. So, why are universities ranked? And, why are they ranked across national borders? The key lies in the movement from a national historical institution to rationalized organization. The general idea is that the more any entity is imagined as a rationalized organization, the more it is at risk of being compared to other entities. Or, to be more precise, the rationalized organization image undermines the historicity or distinctiveness of the entities by dangling portable "best practices" before them. What are then compared are aspects of the entities rather than the entities as wholes. Phrased differently, as long as universities claimed to be deeply rooted in national fabrics and essentially indivisible wholes, they were fairly immune from comparisons. Obviously the older and more successful universities were better buffered from broader standards and inter-university comparisons, not to say international rankings. These universities could turn down proposals for change by simply observing, "things are not done that way here." The invocation of distinctive institutional identity provided the rationale and justification for "why we do what we do." A rich historical "we are" is the sine qua non for "what we do."

Universities as national institutions cannot be compared across national boundaries, but universities qua organizations can be compared with other non-educational organizations. So, for example, the Entrepreneurship Research and Policy Network has recently generated a top ten list of well-managed organizations in which universities such as Harvard, Chicago, Stanford, and Yale make it to the top, along with the National Bureau of Economic Research, the World Bank, the European Corporate Governance Institute, the Centre for Economic Policy Research, and the Institute for the Study of Labor.

Most U.S. universities have neither had the burden nor the advantage of historical richness and local roots. Moreover, they are embedded in a relatively new national landscape. It is therefore not surprising that U.S. universities have always been inclined to think about their "competitors" and to favor inter-university comparisons. An optimistic belief in progress and in science for progress has made U.S. universities inclined to imagine that something is to be learned from inter-university comparisons. Greater faculty mobility across universities is enhanced by this optimistic belief and by relatively low levels of both university historicity and local roots. These factors provide some of the foundation for the rise of the rationalized U.S. university as an organizational ideal.

University rankings within the United States came earlier than the international rankings that now command worldwide attention. Much of this comparative effort predates the highly publicized annual rankings found in the *US News and World Report*, though it is indisputable that they have significantly impacted U.S. higher education (Saunder and Espeland 2009). On the dark side, there is a sense that some places of higher learning are actively working their numbers to show improvement. This may involve an excessive focus on such things as yields (what percent of accepted applicants come to you rather than your competitors) and standards (what percent of applications do you accept). David Kirp (2003) suggests that some admissions offices encourage weak applicants to apply in order to increase the applicant pool. Furthermore, these offices reject worthy applicants who are likely to reject their offers to go to more prestigious universities. The admission offices in earlier "in house" studies have utilized these notions of standards and yields, so this experience gives these measures some face validity. But the public at large was less directly and less dramatically informed as to who were the winners, who were ascending, and who were declining. Thus, the *Report* impacted universities directly and indirectly through increased public awareness of the university rankings. How much the rankings influence student choice is unclear. It is, however, clear that many universities pay attention to the rankings and dedicate time and energy to upgrade their standing.

But the methodological problems with the rankings are known and are a source of skepticism (see, e.g., Welden 2006 on business school rankings). Skepticism abounds but does not stop universities and schools from participating. Efforts to boycott the *Report* have failed mainly because elite universities have not successfully formed a coalition to do so. This e-mail from an Associate Dean for Administrative Affairs is instructive:

> It's the time of year when we send *US News and World Report* our data for their annual rankings of graduate schools of education. Although we don't approve of such rankings and scoff somewhat at their methodology, we have to play the game. One of the questions asks: "How many full time tenured or tenure track faculty have received the following awards or have been an editor of the following journals in the last two calendar years (2006 and 2007)?" These are the awards and journals they include in their list. Please let me know if these apply to you. (Personal communication)

This brief request starts with the assumption that we all understand that this a routine practice. There is a time for giving lectures, presenting papers at professional conferences, attending faculty meetings, and responding to questions from the *Report*. This request comes to us, because now is the time to comply. Notice the requester's awareness of methodological problems, and yet the unproblematic response that "we must play the game."

Why? The answer in part is because much of the information sought is already contained within the annual reports. Awards and journal editorships are already established data entries under the honors and awards and professional activities or services categories in the annual reports. Tenure protocols make use of the same kind of information. Not only are the questions relatively easy to answer but also they are taken for granted features in the everyday life of many U.S. university professors. Methodologically flawed as the rankings may be, the familiarity of the questions lends an aura of legitimacy. It is less than evident in other national systems of higher education, where standardized annual reports are not common practice. In these systems assessments and rankings are often critiqued as features of a rising audit culture that undercuts professorial conviviality.

## Concluding Thoughts

The rationalized university as an organizational ideal emerged earliest in the United States. Both the national and university experiences within the United States made them less historically rich and locally grounded national institutions. Universities in the United States look like "organized anarchies," only because the organizational yardstick was easier to apply to them than to the older and more organizationally hollow but institutionally rich European universities. They have had more experience playing the organizational game, acting as if there were portable lessons to be gleaned from restlessly comparing themselves with other universities. These comparisons presuppose standards and a widespread belief in progress.

But what was it about U.S. universities that moved them in this direction? Joseph Ben-David and Abraham Zloczower (1962) provide the useful insight that, relative to their European counterparts, U.S. universities were less buffered from other social institutions. Neither the authority of the state nor the professoriate shielded the university from a range of pressures emanating from different groups and interests. For the sake of legitimacy and funding, they coped with these different, and at times, competing demands by more carefully managing and properly displaying socially respectably identities. A distinctive and powerful administrative strata emerged in them. Other forms of organizational differentiation would ensue, paving the way for today's distinctive fundraising and public relations organizational units. Thus, an unintended consequence of the university being less differentiated from other institutions is the proclivity for greater organizational differentiation and exercises in accounting for excellence.

Belief in progress and in organizing to attain progress has evolved in two ways, interacting with each other to produce the rationalized university as an organizational ideal. First, the belief is linked to models that attribute success to better organization and management. To the degree that the university is imagined as an organization, the university is more "at risk" of being rationalized. That is, the university is more likely to be attuned to, and receptive of, theories of "best practices" and heroic stories of their embodiments in other universities or systems of higher education. Second, this belief is no longer constrained in its application by national boundaries. Cultural globalization is on the rise and not surprisingly, universities now face world standards of excellence and are pressured to adapt to international benchmarks, increasing both the extensiveness and intensiveness of rationalization.

Of course, transnational models interact with historical legacies. In countries with thicker historical legacies, hybridization or decoupling between formal policy and informal practice are found. In Germany, for instance, "habilitation" status for becoming a professor is no longer legally mandated and junior professorships have been created. But many junior professors write the "second doctoral thesis" for fear that senior faculty will uphold the older standard. However, in most countries universities are fairly new and have neither the resources nor the reputation to rely on historical legacies to maintain themselves.

To be sure, in earlier eras universities have influenced other universities across national boundaries. The influence of the German university in the late nineteenth and early twentieth centuries is well established. But the current globalization of the model of the university differs from earlier patterns of exchange and influence in good part because the state of the globalization of the world differs. First, communications and technological developments have given rise to a world in which models can and do diffuse more rapidly, and these models reach beyond the developed or Western world. This extensive reach facilitates the ease with which higher education is theorized in abstract and universalistic terms, instead of nationally specific idioms that reflect culturally distinctive canons. Second, there has been a worldwide growth of organizational expertise in general, and more recently, of organizational expertise in higher education. This expertise is not historically grounded, but rather is rooted in the triumph of the more ahistorical social sciences and the commitment to the pursuit of progress (Frank and Gabler 2006). This expertise and the theorization it generates are likely to celebrate the entrepreneurial, accountable, and transparent university more than any specific U.S. university. Accounting for excellence practices are hard to resist if couched as rational efforts to gauge quality and progress rather than as distinctive features of an alien national system of higher education. Lastly, the current state of globalization more directly emphasizes the world and global norms, for example, a global ecosystem, common humanity, and the putative equality of all humans (Therborn 2000). The current emphases facilitate the identification and diffusion of "best practices" across an increasing range of domains, including university organization.

To summarize, cross-border diffusion of university forms and practices is not a novel phenomena. What is novel is a world privileging professional expertise that facilitates cross-border diffusion by theorizing portable goals and strategies for attaining these goals and mechanisms for assessing progress toward attainment. What is novel is the technical ease and rapidity of communication flows across national borders. Finally, what is novel is a level of world consciousness shaped in good part by the "scientization of society" and "the socialization of science" (Drori et al. 2003). Nature and society are increasingly imagined as subject to law-like forces that facilitate imagining universities as organizational actors and legitimate structures and activities that communicate transformation.

Skepticism abounds but fails to derail rationalization or undermine the rationalized university as an organizational ideal. Much of this rationalization revolves around the idea of excellence, cast in abstract organizational terms. Universities as distinctive national institutions solely reflecting historical legacies can escape international comparisons, but the ongoing globalization of cultural models of progress makes it increasingly difficult to cling to the imagery of national distinctiveness characteristic of earlier eras. All universities are under varying degrees of pressure to generate accounts of excellence, or at least accounts of commitment to excellence. Many educational reforms throughout the world are better understood if one recognizes how much "accounting for excellence" is a set of practices driven by the broader dynamic of transforming universities into organizational actors.

# References

Ben-David, Joseph, and Abraham Zloczower. 1962. "Universities and Academic Systems in Modern Societies." *European Journal of Sociology* 3 (1): 45–85.

Chabbott, Colette, and Francisco O. Ramirez. 2000. "Development and Education." In *Handbook of Sociology of Education,* ed. M. Hallinan. New York: Plenum.

Clark, Burton. 1972. "The Organizational Saga in Higher Education." *Administrative Science Quarterly* 17 (2): 178–183.

Drori, Gili S., John W. Meyer, Francisco O. Ramirez, and Evan Schofer. 2003. *Science in the Modern World Polity: Institutionalization and Globalization.* Palo Alto: Stanford University Press.

Engwall, Lars. 2008. "Minerva and the Media: Universities Protecting and Promoting Themselves." In *European Universities in Transition,* ed. C. Mazza, P. Quattrone, and A. Riccaboni. Cheltenham, UK: Edward Elgar.

Flexner, Abraham. 1930. *Universities: American, English, and German.* Oxford: Oxford University Press.

Frank, David, and Jay Gabler. 2006. *Reconstructing the University: Worldwide Shifts in Academia in the 20th Century.* Palo Alto: Stanford University Press.

Frank, David, and John W. Meyer. 2007. "University Expansion and the Knowledge Society." *Theory and Society* 36 (4): 287–311.

Gumport, Patricia J. 2000. "Academic Restructuring: Organizational Change and Institutional Imperatives." *Higher Education* 39 (1): 67–91.

Kirp, David. 2003. *Shakespeare, Einstein, and the Bottom Line.* Cambridge, MA: Harvard University Press.

Krücken, Georg, and Frank Meier. 2006. "Turning the University Into an Organizational Actor." In *Globalization and Organization: World Society and Organizational Change,* ed. G. S. Drori, J. W. Meyer, and H. Hwang. Oxford: Oxford University Press.

Meyer, John W., Frank Ramirez, and Evan Schofer. 2007. "Higher Education as an Institution." In *The Sociology of Higher Education: Contributions and Their Contexts,* ed. P. Gumport. Baltimore, MD: Johns Hopkins University.

Musselin, Christine. 2004. *The Long March of French Universities*. New York/London: Routledge and Falmer.

Ramirez, Francisco O. 2006. "The Rationalization of Universities." In *Transnational Governance: Institutional Dynamics of Regulation,* ed. M. Djelic and K. Shalin-Andersson. Cambridge: Cambridge University Press.

Saunder, Michael, and Wendy Nelson Espeland. 2009. "The Discipline of Rankings: Tight Coupling and Organizational Change." *American Sociological Review* 74 (1): 63–82.

Slaughter, Sheila, and Larry Leslie. 1997. *Academic Capitalism: Politics, Policies, and the Entrepreneurial University*. Baltimore, MD: Johns Hopkins University Press.

Soares, Joseph. 1999. *The Decline of Privilege*. Palo Alto: Stanford University Press.

Therborn, Goran. 2000. "Globalizations, Dimensions, Historical Waves, Regional Effects, Normative Governance." *International Sociology* 15 (2): 151–179.

Welden, Linda. 2006. *Ranking Business Schools: Forming Fields, Identities, and Boundaries in International Management Education*. Cheltenham, UK: Edward Elgar.

Wotipka, Christine Min, and Francisco O. Ramirez. 2008. "Women's Studies as a Global Innovation." In *The Worldwide Transformation of Higher Education,* ed. D. P. Baker and A. W. Wiseman. Amsterdam: Elsevier JAI Press.

# Chapter 5

# Quality Assurance and Global Competitiveness in Higher Education

*Isaac Ntshoe and Moeketsi Letseka*

## Introduction

The quality, and quality assurance, movements have become highly contested issues in the advent of new managerialism[1] in higher education. This is because while the notion of quality is critical to institutional autonomy and academic freedom, there are no universal criteria to determine quality in the current conditions of global competitiveness and new managerialism. In this chapter we analyze quality measures and the quality assurance movement in the current global market economy. We investigate ways in which the quality assurance movement has shaped higher education policy and practice and impacted national, regional, and international priorities. The chapter's emphasis is on the following areas: (a) policy borrowing on criteria to measure quality and quality assurance movement, (b) the impact of globalization and new managerialism on perceptions of quality and equity imperatives, (c) institutional ranking and the rating of individual academics as criteria for determining quality in higher education, and (d) implications for policy and theory.

The chapter is organized into four sections. In the first section, we explore ways in which global competitiveness, internationalization and the discourse of new managerialism are shaping notions of quality and quality assurance. In the second section we sketch policies and practices of quality and quality assurance in universities in developing countries.

We argue that these policies and practices are imported and implemented wholesale from developed to developing countries, often with little or no alignment to local contexts. Third, we examine institutional rankings and the rating of individual researchers internationally and in South Africa as an example. In the fourth section, we explore the implications for policy and theory, arguing that self-regulation by higher education institutions (HEIs) remains highly contested under the banner of institutional autonomy and academic freedom.

# Global Tides and Internalization of the Quality Assurance Movement

Our assumption is that the quality assurance movement has been induced by the burgeoning increase in private provision of higher education in nation-states. Furthermore, the quality assurance movement has been an impetus in the demand for higher education, the decline in public funding, and the corresponding cross-border, regional, and transnational provision of higher education services. Driven by global competitiveness discourse, quality assurance is a movement exported from established markets in developed countries, to fast-emerging and less developed economies. Furthermore, policy practice on quality and quality assurance are impacting notions of accountability, institutional autonomy, and academic freedom in higher education (Mok 2000, 2005; Beckmann and Cooper 2004).

Don F. Westerheijden (2003) identifies one attempt to internationalize quality assurance in higher education as the Internationalization Quality Review organized by the European University Association (EUA) in cooperation with the Organisation of Economic Co-operation and Development's (OECD) Institutional Management in Higher Education (IMHE), and the Academic Cooperation Association (ACA). Similarly, the premise of the Global Alliance for Transnational Education (GATE) is that quality assurance needs to be internationalized, and that accreditation agencies need to become active at an international level (Westerheijden 2003).

One of the factors behind the increasing interest in international quality assurance is the acceleration of the globalization of higher education. This is apparent in the mushrooming of higher education provision in other countries, as well as in the increased exchange of faculty and students across the globe (Umemiya 2008). We argue that although countries worldwide have national bodies, with some varying degree of power, to ensure quality in higher education, there are no internationally recognized criteria to determine quality, and quality assurance, notwithstanding the burgeoning proliferation of transnational education.

Despite the lack of a common set of criteria, several notable quality assurance bodies have been created to monitor and evaluate transnational education providers. One such entity is the International Council for Open and Distance Education (ICDE), which is a global nongovernmental organization on quality assurance in education. As a similar body, GATE provides a forum for governments, academics, accreditation agencies, students, and business to discuss and implement quality assurance in transnational education. Another agency, the International Network for Quality Assurance Agency in Higher Education (INQAAHE), which was founded in 1999 in Australia and is recognized by UNESCO, has 63 full time and 40 associate members to promote best practices, including accreditation of qualifications.

Similar regional initiatives have been established to assure quality. These include the Arab Network for Quality Assurance in Higher Education (ANQAHE), the Asian-Pacific Quality Network (APQN), the Caribbean Area Network for Quality Assurance in Tertiary Education (CANQATE), the Network of Central and Eastern Europe Quality Assurance Agencies in Higher Education (CEE), the European Association for Quality Assurance in Higher Education (ENQA) and La Red Iberoamericana para la Acreditación de la Calidad de la Educación Superior (RIACES) in Latin America (Umemiya 2008, 278).

Quality assurance is also an issue in Southern Asian countries where enrollment in higher education has tripled in many cases. While countries in that region have been attempting to establish their own quality assurance systems, the Asian University Network (AUN) has also recently promoted such activities at the regional level (Umemiya 2008, 278). It is worth noting that while initiatives driven by AUN could be seen as a response to the tides of globalization, cross-border higher education service provision and exchange of faculty and students within the region have not been fully promoted, making it difficult to explain why regional quality assurance is so active (Umemiya 2008). More importantly, quality assurance in the Southern Asian region emerged not only as a response to the globalization and internationalization of higher education, but in response to the strong demands created by the regional polices of the Association of Southeast Asian Nations (ASEAN) and the environment surrounding HEIs in the region (Umemiya 2008). This environment is characterized by: (a) exchange of faculty and students, (b) collaborative research activities, (c) information sharing, (d) promotion of ASEAN studies (278).

Another issue pertinent to the discussion is whether or not quality and quality assurance procedures are relative across context. Accordingly, is it desirable to assume that once mechanisms, processes, criteria, or standards are clearly defined and described quality in higher education will be assured?

# Policies and Practices of Quality and Quality Assurance

Policy borrowing of quality assurance measures reflects the following international dimensions: (a) applying internationally agreed-upon criteria, (b) internationalization of the curriculum in the assessment, (c) using international units (programs, institutions) as comparators, and (d) involving international evaluators (Westerheijden 2003, 283).

However, while countries have been trying to create quality assurance systems for their higher education sectors, quality assurance has become an issue beyond individual institutions and countries, due to internationalization and transnational provision—or rather imports and exports—of higher education services (Umemiya 2008). Accordingly, the question of whether there can ever be a worldwide quality assurance label to assure quality assessment and accreditation agencies operating internationally must be addressed (Westerheijden 2003). This phenomenon is part of a global version of an open accreditation system and is supported by an international network of quality assessment agencies.

In the foregoing discussion we have argued that there has been some policy borrowing relating to quality assurance movement across the globe, induced by cross-border provision of higher education. Policy borrowing encompasses methodologies and procedures; criteria for assuring the quality of institutions, programs, governance and management; and the responsiveness of the higher education sector. Policy borrowing is also evident in the criteria for institutional ranking and the ranking of individual faculty.

# Ranking of Institutions and Rating Faculty: Criteria for Assuring Quality

## International Trends

The ranking of universities is a common phenomenon in many Western countries, including the United States, the United Kingdom, Australia,

New Zealand, and in many Asian countries, including China and Japan. Key aspects of rankings include the following: student head count, research productivity, faculty qualifications, international reputation of institutions, responsiveness to market demands, study programs, throughput rate, library holdings, and the quality of teaching.

Though institutional ranking has been a common practice for some time, it has gained popularity with increasing marketization of higher education, greater mobility of students, and ultimately, recruitment, and competition for foreign students (Harvey 2008a). Harvey (2008a) argues that institutions often use their ranking in marketing and promoting themselves while students also use rankings to choose institutions. For instance, the reputation of Monash University in Australia among international students was based on ranking (AUQA 2006; Thakur 2007).

Apparently, ranking has several purposes. First, rankings respond to demands from consumers for easily interpretable information on the standing of HEIs, satisfying the need for consumer guidance among the student population. Second, rankings have become a basis for accountability in the higher education sector and provide a rationale for allocation of funds. Third, rankings stimulate institutional competition. Fourth, they assist in distinguishing institutional types, programs and disciplines. Fifth, rankings serve as a framework to determine quality of higher education, to complement work conducted in the context of quality assessment and review performed by public and independent accrediting agencies (CHEDC et al. 2006; see also Harvey 2008a).

Elements of policy borrowing are evident in the quality assurance movement because methodologies and procedures are usually derived from powerful and fast-emerging economies when ranking institutions. Liu and Cheng (2005) cast doubt on ranking criteria and weights used to select institutions, arguing that institutions are generally ranked according to their academic or research performance. Ranking indicators include alumni and staff winning Nobel prizes and fields medals; highly cited science articles indexed by the Science Citation Index-Expanded (SCIE) and the Social Sciences Citation Index (SSCI); and the size and academic performance of an institution (Liu and Cheng 2005). More revealing about the politics of ranking is that only publications in SCIE and SSCI are considered in the world ranking exercise.

Thus, world university rankings, such as the Shanghai index, are dominated by wealthy countries, especially the United States, the United Kingdom, Japan, Germany, Canada and France. This preeminence is not unexpected, given that developed countries dominate and control conceptions of knowledge, how it is produced and disseminated. In the 2008 Shanghai index, North and Latin America account for 85 percent of the top 20 institutions and 55 percent of the top 100 institutions. The United States has 17 universities out of the 20 top world universities, followed by the United Kingdom with two and Japan with one. Furthermore, the United States has 54 universities in the top 100 universities in the world, 117 out of the top 300, and 166 out of the top 500 world universities. The United Kingdom is second with six universities in the top 100, 33 in the top 300, and 42 in the top 500. Thus, most of the top universities are in developed countries, highlighting the centrality of economic power in building world-class universities (Liu and Cheng 2005).

What is also important to note is that rankings have now become a rule and not an exception to the extent that developed countries will rank universities regardless of whether a country has a ranking practice or not. Commenting on Russian higher education, Sadlack (2006) notes that the global dimension of university ranking is confirmed by the fact that if an institution does not produce its own ranking it should not be surprised that others are going to do so for it. For example, the Huazhong University of Science and Technology in Wuhan, China, published a ranking of Russia's top 100 universities.

## The South African Case

South Africa is not immune to the influence of the quality assurance movement following the country's readmission to the international community after the demise of apartheid. As in other countries, emphasis on quality assurance was induced by globalization and internationalization. Thus, the Higher Education Quality Committee (HEQC) in South Africa has established and maintained relationships with quality assurance bodies and organizations in Africa and internationally, especially systems in countries with powerful economies (CHE 2008). Understandably, quality assurance in South Africa has become necessary given increased private provision through cross-border and transnational higher education services the country has experienced since the 1994 democratic elections. These cross-border and transnational provisions have necessitated the development and implementation of frameworks, criteria and procedures for quality assurance programs to protect students against rogue providers.

A key feature of the quality assurance discourse in South Africa is the absence of institutional rankings within the country as an official policy to determine quality (CHE 2002). Despite this, there has been evidence of policy borrowing driven by internationalization and globalization to compare South African institutions with those in developed economies and less developed economies.

According to one study, many South Africa universities compete well with some of the institutions in wealthy countries and all middle-income and developing countries (Study South Africa 2009). Unsurprisingly, they dominate the African ranking of universities as well. In 2008, South Africa's highest ranked university—the University of Cape Town (UCT)—was ranked 179th in the world by the *Times Higher Education* (*THE*), ahead of the University of Colorado, which was ranked 180, Lomonosov Moscow State University, which was ranked 183, and the University of Massachusetts, Amherst, which was ranked 191.

Anastassios Pouris (2007) explains that citation of articles is often used for evaluation purposes. He identifies South African universities that have been included in the top one percent of the United States' Essential Science Indicators (ESI) database in the past ten years, across 22 scientific fields, namely the universities of Cape Town, Pretoria, Free State, the Witwatersrand, KwaZulu-Natal, and Stellenbosch. Indeed, South African universities also dominate the African universities' rankings with regard to publications. Ten South African universities are in the top 12 positions of the Webometrics ratings, and are separated by one university each from Egypt and the island of Reunion (Webometrics 2009). Furthermore, 12 South Africa universities are in the top 25 of the Webometrics rankings, with three from Egypt, two from Kenya and one each from Reunion, Tanzania, Zimbabwe, Senegal, Namibia, Mauritius, Morocco, and Mozambique. The domination of South African universities in the continental rankings mirrors the country's continental economic dominance and the extent to which it has embraced internationalization and globalization.

Notwithstanding our reservations about rankings, they are important to students, research administrators, industry, and academics. According to Pouris (2007), rankings may be used as a proxy for employment opportunities because they serve as screening devices and indicators of research quality for employers, who compete for graduate students from reputable academic institutions and offer positions well in advance of the students' graduation year.

It should be noted that despite the absence of formal ranking of institutions within South Africa, institutions are ranked, albeit subtly, in terms of research output, the number of rated academics, and the extent to which institutions secure third-stream income from industry. Furthermore, South Africa's "big five"—Witwatersrand (Wits), Pretoria, Stellenbosch, KwaZulu-Natal, and UCT, which were formerly White, advantaged universities with an established culture and track record of research—are cited in many world rankings of universities.

Publications in the Science Citation Index-Expanded and the Social Sciences Citation Index are the most common criteria used in South Africa to rank institutions (Pouris 2007). Universities that offer programs in the natural sciences, technology, and medicine are usually ranked highest. Six South African universities that offer clinical medicine, plant and animal science—UCT, University of Pretoria, Orange Free State University, Wits, University of Natal, and University of Stellenbosch—appear in the ESI database in recognition of their research character (Pouris 2007). Three universities (UCT, Pretoria, and Kwa-Zulu-Natal) were listed in the fields of environment and ecology, three (Cape Town, Wits, and KwaZulu-Natal) in the social sciences, Pretoria and Wits in engineering, and Cape Town and Wits in geosciences. UCT was the only university that met the threshold in biology and biochemistry, while only Wits was listed for chemistry and materials sciences (Study South Africa 2009). Conspicuously absent in the South Africa rankings are the historically Black and disadvantaged institutions, which, for historical and political reasons, were created as teaching institutions offering mostly humanities and social sciences.

Harvey (2008b) argues that university rankings are usually based on a single indicator that derives from a set of indicators that combine into a single index. He agrees with Council on Higher Education Development Chair et al. (2006) that rankings are adopted for convenience and that they are intended to stimulate institutional competition, differentiate types of institutions and programs, contribute to national evaluations, and contribute to debates about conceptions of "quality" in higher education.

For our purpose in this chapter, Harvey's (2008b) four features of rankings are worth noting. First, there is little evidence to suggest that the selection of indicators to rank institutions involves any reflection. It does seem, though, that rankings are based on convenience, where the quality of teaching and other factors are difficult to establish. Second, there is no evidence to suggest that the weightings are theoretically grounded. Rather, they are simply arbitrary. Third, publication of rankings does not accurately reflect real changes in institutions that are rarely noticeable within one year. And fourth, little consideration is given to institutional, political, and cultural environments that affect how institutions operate and what they can do.

Rankings are also problematic for institutions in the developing world, most of which do not have sufficient research capability (Harvey 2008a). Furthermore, Thakur (2007) notes that the ranking systems have had an impact on HEIs and their stakeholders, indicating that one of the top universities in Malaysia, the University of Malaya, dropped 80 places in the *THE* rankings without any decline in its real performance, due to definitional changes.

For Clark (2007), rankings might have a negative impact on equity. This is because often student selection indicators used in some university rankings, such as test scores for entering students and the percentage of applicants that institutions accept, threaten higher education access for disadvantaged students by creating incentives for schools to recruit students who will be "assets" in so far as maintaining their position in the rankings is concerned. Thus, in order to improve their performance on these measures, institutions engage in various strategic activities to compete more effectively for academically high-achieving students, which tend to have a negative impact on access for low-income students and other underrepresented groups (Harvey 2008a).

More seriously, Harvey (2008a) correctly notes that the Shanghai rankings from China have acquired an existence beyond their purposes as they are now used by some as a ranking hierarchy of institutions worldwide, while the basis of their compilation is often overlooked. Institutions are ranked according to their academic or research performance:

> Ranking indicators include alumni and staff winning noble Prizes and fields medals, highly cited researchers in twenty-one broad subjects categories, articles published in *Nature* and *Science*, articles indexed in Science Citation Index-Expanded (SCIE) and Social Science Citation Index (SSCI), and academic performance with respect to the size of an institution. (Harvey 2008a, 198)

These criteria and procedures often do not reflect the geographical, national, regional, institutional, and political environment within which quality assurance is measured (see Harvey 2008a). Accordingly, despite the differing contexts that make international comparison theoretically complex, international ranking lists barely take contextual information into account (Harvey 2008a).

In South Africa, while institutional ranking is not a policy, there is a practice to rank individual academics. This is not peculiar to South Africa, and represents policy borrowing from other systems, including New Zealand and Mexico (Pouris 2007). Accordingly, procedures and methodologies used for rating individual academics and researchers in South Africa are borrowed from other countries and are part of the broader global quality assurance movement. We want to argue, though, that the National Research Foundation's (NRF) rating of individual academics and researchers is biased toward historically White and advantaged institutions. It reproduces institutional inequalities and inequities and reflects unequal power relationships in the country's higher education sector.

The methodology, procedures, and criteria of rating individual academics and researchers are imported from developed countries and powerful economies including a preference for articles published in the ESI database of the Institute for Scientific Information (ISI) (Pouris 2007). Reflecting policy borrowing further, historically White institutions offering natural sciences and medicine have the highest number of rated academics, but also have the highest rated academics in the humanities and social sciences. We endorse Michael Cherry's (2008) observations that the process of academic ratings has two central limitations. First, it rewards researchers who publish in a particular field, thereby penalizing work that is multi-disciplinary. Second, there is ambiguity regarding what constitutes "international recognition" and a "proven track record."

Problematic too is that procedures and criteria for ranking in South Africa are derived from natural and earth sciences fields, and this tends to disadvantage the rating of academics in human and social sciences. Our view is that this emphasis on the hard sciences has the potential to undermine interdisciplinary, trans-disciplinary, and multi-disciplinary work, inherent in human and social sciences.

## Implications for Policy and Theory

The foregoing discussions suggest policy borrowing in the quality assurance movement. Our view is that this process reflects global economic developments engendered by cross-border provision of higher education by transnational foreign providers. This tends to undermine regional and local conditions. Embedded in the global competition and market ideology that favor developed economies, quality assurance is determined by the number of articles published in reputable journals, production of measurable outputs, provision of so-called viable programs, industry-university partnership in innovation, success in raising second-stream income, and students' completion rates.

This trend has policy and theory implications that warrant further discussion. First, current notions of the global quality assurance movement now transcend national boundaries and have the effect of undermining regional and national autonomy and sovereignty. While the development and usage of universal criteria and procedures to assure quality in higher education have become unavoidable, they are necessary given the widely accepted practice of cross-border provision of higher educational services. For example, under the Bologna Process, it is expected that universities in Europe will cooperate in the establishment of a common framework of reference, to disseminate best practices as well as harmonize the policies of individual countries (Westerheijden 2003).

Second, policy makers must be wary of wholesale borrowing of policy and the practice of ranking institutions from developed and fast emerging nations and regions imported to less economically powerful economies without, or with little, adaptation regardless of the diverse contexts of countries. At the macro level, this borrowing is likely to overlook macro-economic policies of countries and national imperatives such as address-ing institutional equality and equity. Thus, an imported ranking policy may undermine a soundness of "fitness for purpose" that presupposes that institutions have different purposes and should be judged against those criteria while rankings establish and judge against a set of generic criteria (Harvey 2008a).

Third, it is instructive to be wary of the fact that institutional rank-ing is clearly biased toward institutions that have sometimes uncritically embraced marketization, commodification of knowledge, and commer-cialization and entrepreneurship to the extent that they are able to raise additional income from industry. There is also a great concern in South Africa that even though institutional ranking is not a policy, current fund-ing policy favors institutions that enjoy international recognition.

## Conclusion

We have argued that the quality assurance movement is an example of policy borrowing exported wholesale by developed economies. We under-scored that regardless of whether developing countries may or may not have institutional ranking as policy, their HEIs will be ranked by international ranking agencies due to unequal power relations and global competition. This movement indirectly undermines regional and local conditions. We argued that institutional ranking and the rating of individual academ-ics are typical examples of policy borrowing that have become acceptable in the current context of global competition. Thus, the quality assurance movement has become an economic ideology that has little to do with assuring quality but everything to do with the imposition of foreign crite-ria, standards, and methodologies under the guise of harmonization and alignment of the policies to facilitate global competitiveness.

## Note

1. We borrowed the term *new managerialism* from Rosemary Deem (2001) to denote the permeation of values and ethos of business into higher education. As part of the new managerialism, higher education is increasingly required to demonstrate the same kind of accountability demanded by corporate business.

# References

Australian Universities Quality Agency. 2006. *Report of an Audit of Monash University*. Melbourne: Australian Universities Quality Agency.

Beckmann, Andrea, and Charlie Cooper. 2004. "Globalization, the New Managerialism and Education: Rethinking the Purpose of Education in Britain." *Journal for Critical Education Policy Studies* 2 (2): 1–32.

Center for Higher Education Development Chair, Center for Education, Institute for Higher Education Policy. 2006. *Berlin Principles on Ranking of Higher Education Institutions*. Berlin, May 20. http://www.che.de.

Cherry, Michael I. 2008. "Time to Stop Rating Researchers." *Mail and Guardian: Higher Learning Supplement*, June 30: 3.

Clark, Marguerite. 2007. "The Impact of Higher Education Rankings on Student Access, Choice, and Opportunity." *Higher Education in Europe* 32 (1): 59–70.

Council on Higher Education. 2002. "Higher Education Quality Committee. Programme Audit Framework." Draft Document. Pretoria: Council on Higher Education.

———. 2008. *Coordinating Quality Assurance in Higher Education*. Pretoria: Council on Higher Education.

Deem, Rosemary. 2001. "Globalization, New Managerialism, Academic Capitalism and Entrepreneurialism in Universities: Is the Local Dimension Still Important?" *Comparative Education* 37 (1): 7–20.

Harvey, Lee. 2008a. "Ranking of Higher Education Institutions. A Critical Review." *Quality in Higher Education* 14 (3): 187–207.

———. 2008b. "Assaying Improvement." Paper presented at the 30th Annual EAIR Forum, Copenhagen, August 24–27, 2008.

Liu, Nian Cai, and Ying Cheng. 2005. "Academic Rankings of World Universities: Methodologies and Problems." *Higher Education in Europe* 30 (2): 127–136.

Mok, Ka-Ho. 2000. "Impact of Globalization: A Study of Quality Assurance Systems of Higher Education in Hong Kong and Singapore." *Comparative Education Review* 44 (2): 148–74.

———. 2005. "The Quest for World Class University: Quality Assurance and International Benchmarking in Hong Kong." *Quality Assurance in Education* 13 (4): 277–304.

Pouris, Anastassios. 2007. "The Institutional Performance of the South African Institutions: A Critical Assessment." *Higher Education* 54 (4): 501–509.

Sadlack, Jan. 2006. "Validity of University Rankings and Its Ascending Impact on Higher Education in Europe." In *Bridges* 12, December 14. Washington, DC: Office of Science and Technology (OST). http://www.ostina.org.

Study South Africa. 2009. *University Rankings—South Africa Holds Its Own*. Pretoria: International Education Association of South Africa, Study South Africa. http://www.studysa.co.za/.

Thakur, Marian. 2007. "The Impact of Ranking Systems on Higher Education and its Stakeholders." *Journal of Institutional Research* 13 (1): 83–96.

Umemiya, Naoki. 2008. "Regional Quality Assurance Activity in Higher Education in Southeast Asia: Its Characteristics and Driving Forces." *Quality in Higher Education* 14 (3): 277–290.

Webometrics.com. 2009. *Ranking Web of World Universities.* Madrid: Webometrics. http://www.webometrics.info.

Westerheijden, Don F. 2003. "Accreditation in Western Europe: Adequate Reactions to Bologna Declaration and the General Agreement on Trade in Services?" *Journal of Studies in International Education* 7 (3): 277–302.

# Chapter 6

## Quality-Oriented Management of Higher Education in Argentina

*Héctor R. Gertel and Alejandro D. Jacobo*

### Introduction

Globalization is a central concern in higher education. Globalization in education implies the recognition that investment in knowledge acquisition and dissemination is not only necessary for keeping high standards of economic and social well-being but also essential in times of economic downturns. Global research competition affects academic output of individual countries, transforming the labor market for science and the professions, and promoting improvements in teaching (Hicks 2007). Globalization in general and in higher education are intricately linked. Within this paradigm, the challenge higher education institutions (HEIs) are facing today is how to timely react to globalization, and how to accommodate themselves to massive demands while still assuring the delivery of quality teaching and research (Pérez Lindo 1998, 51). As a consequence, many countries have begun to implement innovative procedures for higher education quality assurance.

In countries that lead in technology innovation, university management contributions to quality assurance concentrate on accounting, administrative, and audit procedures. These procedures generate data and measures to evaluate teaching and research impacts on teaching, thus enhancing the global competitiveness of higher education. However, in countries characterized as technology followers—including Argentina—higher education

management has been slow in aligning itself with the external requirements of quality assurance (Gertel and Jacobo 2007a, 2007b). The reasons
these universities have been slow to react are not completely clear, and we
explore some potential causes in this chapter.

This chapter concentrates on the adjustments the national universities
of Argentina have made with regard to quality assurance. These institutions represent well over 80 percent of total enrollments and 90 percent
of full-time equivalent teaching staff. Almost all of the university sponsored research takes place at these institutions; private universities capture the remainder of enrollments and contribute marginally to research.
Our underlying hypothesis is that Argentina's slow response results from
the complexity associated with the decision-making process of collegiate
governing bodies, as has been previously suggested by a recent assessment
of Argentina's strategy of human capital formation (Holm-Nielsen and
Hansen 2003).

> Within the limits of this chapter, quality is considered to be the result of
> a strategic alignment of the internal decisions on authority, financing, and
> evaluation with the external demands placed by the government and the
> global society. The relevance of strategic alignment fits into the idea of
> quality as the outcome of mutual collaboration and of effective information
> systems implemented in the service of stakeholders. In this sense, align-
> ment is the recognition that different sets of decisions concerned with the
> achievement of goals are often taken at different times, at various levels and
> by different people within the organization. (Luftman 1996; Pirani and
> Salaway 2004)

The debate over quality in teaching is strongly influenced within the universities by the conflicting perceptions stakeholders—professors, alumni,
and students—hold about the relationship between science and professional training and about their choice of specific actions and managerial procedures for making scientific activity results accountable. These views are
not necessarily tuned to government or global competition requirements.

# Globalization and Higher Education Management

Globalization is a highly elusive term commonly used to explain the
increasing presence of global networking. Among its many accepted uses,
globalization is associated with the fact that events taking place in one
region of the globe will cause a variety of complex effects on the life and

welfare of individuals and institutions across nations worldwide, including university academic organizations.

The effects of globalization are best understood in terms of three sets of simultaneous contradictions: convergence and divergence, inclusion and exclusion, and centralization and decentralization (Jones and Fleming 2003). In addition, economists subscribe to the idea that globalization comes in waves (Jones 1998; Weil 2005). In their view, technological innovations in transportation and communication during the last quarter of the nineteenth century have spurred contemporary globalization. These innovations have caused the costs of international freight and communications to drop dramatically, and have dynamically evolved into a technology divide that, today, seems to separate countries into convergence clubs (Barro and Sala-i-Martin 2005). Membership to the first club is limited to those countries that led innovation during the past several decades and aligned their institutions—including higher education—toward that end. The second consists of non-industrialized countries that behave as technology followers, whose chances to buy a membership ticket into the first club are real, but contingent on the breadth of the technology gap and on how motivated they are to align their human resources and institutional goals toward a sustainable construction of knowledge capital.

In the late nineteenth century, Argentina, together with Canada, Australia, and New Zealand, was among the group of nations outside those of the industrialized world, who experienced the effects of networking on a global scale and introduced institutional innovations, including the reorganization of leading universities (Halperin Donghi 1962), to reduce the technology gap. Since those early attempts, not only has Argentina been opened to trade in global markets, but it has also simultaneously been receptive to flows of immigrants and to new ideas and technologies. These have contributed to the process of constructing knowledge capital in manufacturing, export-oriented sectors, and government; however, tertiary education management does not seem to have been influenced by this wave of renovation (Waisman 1987).

# Patterns of Authority, Financing, and Evaluation

Around the world, institutions of higher education may be centrally operated by a national government applying a top-down managerial approach, or their decisions may be the result of a decentralized bottom-up democratic process (Tsai and Beverton 2007). In the case of its universities,

Argentina chose a mixed system in 1885 that combines a top-down centralized regulatory framework with university autonomy in academic matters, which, since then, has contributed to their independence from government policies and shaped specific patterns of authority, finance, and evaluation.

In terms of authority, higher education generally enjoys great autonomy or self-government, and its distribution is mostly in the hands of professors (Clark 1983). The national universities in Argentina have multi-party governing bodies in which the different stakeholders have a voice and voting power that are not necessarily aligned to quality assurance. For instance, faculty members holding a doctoral degree, those who would potentially promote quality standards, represent only about 12 percent of the total, and are thus almost voiceless in university government.

In terms of financing, most countries subsidize non-compulsory education, and, more specifically, university education. Arguments in favor of this policy are based on capital markets' imperfections, which prevent students from borrowing against future human capital income, and externalities, associated with scientific production. The traditional system of assigning block-grant subsidies still dominates in many countries (Williams 1990; Strehl et al. 2007). Two new forms of financing were introduced around 1980: selectivity in public funding and expansion of other sources of funds. Because earmarked funds may represent a sizable proportion of a public university's total budget, the introduction of the new forms has trimmed off university authority on resource allocation. Today, in Argentina's national universities block grant subsidies cover the total payroll of faculty and non-academic staff and basic services, while specific monetary incentives, research funding, and building construction and research facilities are increasingly dependent on specific earmarked funds made available by the government, not without tough negotiations on the quality standards to be met (Piffano 2007).

In terms of evaluation, although universities have long implemented self-assessment and auditing mechanisms, external evaluations are more recent. The introduction of external evaluation is usually followed by the enactment of a regulatory framework set forth by the national government. This regulatory framework typically stresses the need for, and encourages the development and implementation of, "objective" standards, methodologies, and research—in other words, benchmarking—which have to do with the management and assurance of quality of the undergraduate and graduate education programs (Middaugh 2001). Universities in Argentina are responding slowly to benchmarking and, as we shall see, this is a historical problem of alignment that conditions the current patterns of higher education quality management.

# Past Trends

The successive governments of Argentina have attempted to align the universities with the principles of positive and experimental science required in the global context. First, a Napoleonic, or centralized, model was enthusiastically backed by the state soon after independence in a process aptly described by Andrés Bernasconi (2008) as intended to "train the professional, secular elites, especially civil servants, in whose hands the building of the new republics was entrusted" (27).

As mentioned above, by the turn of the nineteenth century, Argentina became exposed to, and reacted to, the global requirements of the time. The country received an important flow of immigrants who were the source for the rapid and low-cost increase in the supply of well-trained scientists (Lewis 1978). However, the national universities were slow in responding to these challenges because they were not only reluctant to incorporate foreign fellows as faculty members, but also because they had no previous experience in developing new areas of knowledge. Again, the state understood the opportunities provided by integration into the global society, and took the lead to expand and staff science and engineering teaching and research areas at the national universities. Autonomy was granted to generate self-government in the hands of the faculty, when the *Avellaneda Act* of 1885 was sanctioned. Yet, the newly sanctioned autonomy was unable to counteract nepotism, patronage, and the promotion of professors on the basis of seniority rather than merit and achievement. As a result, most of those immigrants trained in scientific positivism ended up forming the teaching staff of the normal schools while universities lost the opportunity to boost quality teaching (Weinberg 1995).

The question of inclusiveness and quality education was re-introduced at the national universities with more virulence by the student revolt of 1918, known as "the Reform" (Del Mazo 1957). The reformist values were adopted by the national universities of Argentina, and multi-party governments were established, with participation of faculty, students, and alumni and, later, non-academic staff as well (Bernasconi 2008). Decentralized management focused on how to effectively develop and apply administrative systems that would give support to the promotion of positivist disciplines, innovative methods of teaching, and original research. This rule, based on multi-party governing bodies, which differs markedly from the self-governing bodies in the hands of professors commonly found in other countries, still dominates national universities' management organization in Argentina. The accent of the new model was put on issues of access and openness (Schugurensky 2002), efficiency, academic quality, and research.

However, the many critical voices that expressed discontent with the state of affairs must be taken as evidence of slowness of response to important organizational deficiencies inherited from the past that were not solved by the democratic multi-party government (Waismann 1966).

# Current Patterns

## Authority

Understanding how universities are governed in Argentina is essential for assessing the speed of adjustment to global external requirements. In Argentina, all universities are regulated under Article 29 of the Higher Education Act of 1995, which reinforces the traditional concept of autonomy (Ministerio de Educación 1995). Universities are organized into schools, each run by a Dean elected every three years by the academic council, a collegiate body of elected faculty members, students, alumni, and non-academic staff. The University Assembly, an integrated body of representatives from all academic units, elects the Rector of the University every four years. The Rector's initiatives, however, have to be validated by a different legislative board of governors, called Consejo Superior, formed in turn by all the heads of academic units, plus a number of elected members representing the faculty, students, alumni, and in some cases, non-academic staff. A salient feature of the Consejo Superior, whose members often share sympathies with national or local political parties, is its conflicting views on how to respond to quality standards for research and education. Authority is consequently diluted and the non-academic staff members perceive an atmosphere where the sense of direction is often hard to envisage. As a result, the universities generally do not reach the alignment expected by the government and the civil society, which is required to supply quality in education and research. Under the current system, the academic councils do not always approve professors' initiatives, since professors alone do not have the majority vote needed for their approval. Thus, the final balance in decision-making depends on coalitions that not only introduce confusion and uncertainty in university governance but also hinder the adoption of tactical decisions required to meet quality standards. This is what we define as a slow response.

## Financing

National universities are completely supply-funded by the national government and, although they are allowed to collect money from extra sources, fundraising activities are not part of their culture (Ministerio de Educación 1995, Section 28). They are even legally allowed to charge tuition fees, but this prerogative has never been applied. The Consejo Superior is responsible for the distribution of the university budget.

The national government supplies most of the funds in the form of a block grant. Traditionally, this block grant for teaching and research has been allocated through a mechanism based on the institution's previous share of the budget and its lobbying activity in Congress (Fanelli 2007; Piffano 2007). Since 1992, new mechanisms have been introduced to fund national universities in order to improve efficiency and equality by means of specific grants. The higher education authority distributes some funds, while additional funds come from the science and technology authority.

In 1997, the higher education authority introduced a new formula to allocate a small but increasing proportion of earmarked public funds to national universities (Delfino and Gertel 1996). At the beginning, the formula was not easily accepted by the universities because it meant a departure from the traditional block-grant funding mechanism that still represents the largest source of income.

A short-lived, but nonetheless important, earmarked fund was implemented between 1995 and 2000 to enhance institutional quality. To this end, the higher education authority allocated funds on a competitive basis via a program called Fondo para el Mejoramiento de la Calidad Universitaria (FOMEC). FOMEC was implemented as a competitive funding mechanism to select institutional projects on the basis of merit and pertinence according to their quality. Over 1,400 scholarships for graduate studies were issued by FOMEC, but scarcely 20 percent of the graduates returned to the national universities. One of the main reasons for this low rate of attraction can be associated with the unwillingness of the universities' collegiate governing bodies to review the traditional hiring procedures, based on seniority rather than on academic credentials. This practice negatively affected rotation of faculty and, consequently, academic quality. Coupled to FOMEC, a fund was created to supplement salaries to faculty members who both teach and conduct research—less than 18 percent of the total in 1998. However, the relative value of this supplement decreased over time and discouraged newcomers to the system. Faculty participation in the program decreased to 15 percent of the total by 2006 (Ministerio de Educación 2008).

In addition to these earmarked funds granted by the higher education authority, the growing importance of the science and technology authority as the leading provider of funding for scientific research carried on at the national universities should be mentioned. As a result of this government initiative, the share of national universities in the science and technology national budget decreased from approximately one third of the total budget in the 1990s to an average of one fourth for the last five years.

The recent upgrading of the Argentine authority for scientific affairs to full ministerial rank underlines the great significance assigned to knowledge and innovation for Argentina's future and further reinforces the tendency of the state to be involved in the enhancement of academic and institutional quality in the national universities. The new Ministerio de Ciencia, Tecnología e Innovación Productiva (Ministry of Science, Technology and Innovation in Production, or MINCYT), has already expressed the intention to overhaul the currently fragmented science and technology system in Argentina by putting greater emphasis on multidisciplinarity and flagship initiatives.

Thus, today, an increasing amount of the national universities' budget that should guarantee excellence in teaching—by reinforcing quality research and networking, a pillar of the reformist movement—comes paradoxically, not from supply-driven university funds, but from demand-driven research projects. These projects are generated and managed by individual research groups, and more recently, by networking national universities' research groups, with financing mainly provided by the MINCYT, further eroding the authority of the national universities' collegiate bodies in matters of how scientific research should be conducted, as well as in decisions concerned with the direction and quality assurance of teaching in graduate schools.

## Evaluation

External evaluation of universities was introduced in 1996 in Argentina, when the *Higher Education Act* created the Comisión Nacional de Evaluación y Acreditación Universitaria (National Committee of University Assessment and Accrediting, or CONEAU) to foster quality improvements. This committee is in charge of universities' external evaluation and accreditation of government regulated undergraduate and graduate programs, among other functions.

CONEAU is entitled to grant accreditation to undergraduate programs issuing a first degree for government regulated professions. The implementation of these exercises is relatively recent and data to produce comparative research is not yet available. In addition, as has been mentioned, CONEAU is in charge of the regular accreditation of graduate school courses. CONEAU has also produced an important set of tools to conduct the evaluation and accreditation process, including guidelines for data collection. The guidelines comprise sections with focal points of analysis intended to verify compliance with the standards.

External evaluations conducted by CONEAU every six years are complementary to self-evaluations performed by the institutions in the analysis of their own achievements and difficulties. Although external evaluation reports are public, guidelines for performing external evaluation are merely indicative, and the resulting reports are consequently neither uniform nor comparable.

In its short life, CONEAU has produced external evaluation reports for 22 national universities out of a total of 37. Analysis of authority, financing, and audit strengths and weaknesses reported by peer evaluators was conducted using a random sample of 30 percent of the national universities' reports that were approved. Authority was found to be fragmentary and weak, and attributed to prevailing strong political distrust among members of the multi-party governing bodies, a feature documented by all the reports. Authority, most reports observed, was further weakened because the organization of the institutions resembled a federation of small kingdoms—each professional school having little in common with the others. Financing is said to be scarce and it is distributed among academic units on the basis of historical trends. However, all the reports emphasized that the most important finding was a painful lack of interest in improving objective indicators of performance—drop-out rates, duration of a career, costs, enrolments—that would help improve efficiency. Because of weak authority and scarce control in the use of funds, most peers indicated that imbalances have developed in the distribution of non-academic personnel between the central administration and the academic units, and within the academic units between the Office of the Dean and the academic departments. Furthermore, funds for research are obtained from external sources by individual research groups and are beyond the control of the university. In short, organizational fragmentation is reflected in strong weaknesses in—in other words, absence of—supportive information systems that undermine the possibilities of most institutions to establish modern accounting, administrative, and audit systems.

Between 1997 and 2006 CONEAU received 1,435 requests for accreditation of graduate programs of which a total of 1,116 requests submitted by national universities were processed. Only 48 percent of accredited programs received an excellent or very good grade. The percentage was higher in Basic Sciences (84 percent) and Applied Sciences (60 percent), followed by Health Sciences (44 percent). The lowest values were found in Humanities (35 percent) and in the Social Sciences (33 percent).

The number of requests submitted to CONEAU equals half of the graduate degree programs offered by Argentine public and private universities. This suggests that one out of every two programs voluntarily submits a request, and marks a clear trend toward an improvement—or alignment—of the quality of the programs. Slowly, the universities are opening up, and inserting themselves into the challenges posed by globalization. However, although evaluation has shown a favorable evolution, evaluation and funding are not yet strictly and formally linked to each other. In the case of undergraduate programs, the accreditation of careers is embryonic. In the case of graduate programs, outstanding accredited programs have benefited indirectly, by capturing students with fellowships from the national research authority, known as Consejo Nacional de Investigaciones Científicas y Técnicas (CONICET), and students from abroad. Thus, a university that has not achieved alignment voluntarily is forced by external forces to do so. Such forces may be external accreditation and quality assurance evaluation agents, and external research funding institutions.

# Assessing the Quality of National Teaching and Research

The interest for higher education quality management is rather new in Latin America in general, and in Argentina in particular. It is an expression of growing claims from the different social actors and generates concern among those involved in the governing bodies of universities.

Student unions perceive threats from global shocks and external government intervention through demand-driven incentives because they are heavily affecting the traditional self-governing model of the national universities. Professors feel the multiplication of professional programs is negatively affecting the conditions of research and quality teaching. Graduates are demanding continuous education training courses that compete with the traditional programs offered. This set of concerns has helped to introduce a process of reflection and adjustment in the national universities, which is still underway.

The guidelines provided by CONEAU include nearly all the indicators required for benchmarking in teaching and research suggested in the literature. However, the alignment under the CONEAU standards—benchmarks—involves a technical dimension and a social one. Due to the composition of collegiate governing bodies, internal divisions as to how to interpret the value of external evaluation and on how it will impact autonomy are increasingly visible. This means that the new social context of globalization remains an area of conflict. Despite their resistance to external quality evaluation, national universities have made progress toward quality that, arguably, is the result of the work of CONEAU.

As an example, progress toward quality assurance can be measured in terms of external indicators such as the Science Citation Index (SCI), which registers the documented involvement by researchers in various science and technology publications. This data source is particularly useful in the context of this chapter as an indicator to check the advances made by the national universities toward quality.

Figure 6.1 summarizes the share of the institutions in the number of published scientific articles registered in the SCI. National universities and CONICET are grouped together since researchers normally belong to both institutions, while the rest of publications mainly come from private universities.

Figure 6.1 shows the significance of publications made by CONICET and the national universities research groups that have increased their share from 83 percent in 2000 to 86 percent in 2006. This means that during this period only one out of five publications came from private universities.

Figure 6.2 summarizes the scientific production by type of publication according to the SCI classification in scientific papers, abstracts, review articles and others.

The share of scientific papers has been relatively stable between 1990 and 2006 at a level of roughly 85 percent. However, the numbers of abstracts

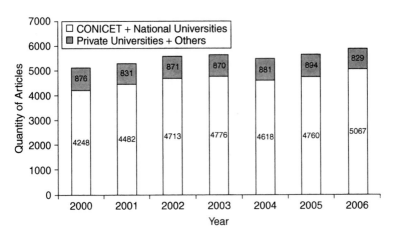

**Figure 6.1**    Scientific Articles by Type of Institutional Affiliation
*Sources*: CAICYT (2008) and MCTIP (2008).

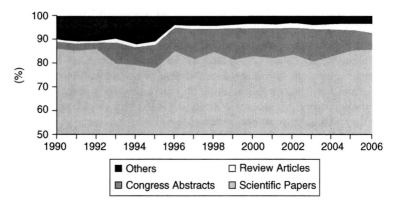

**Figure 6.2**    Scientific Production by Type of Publication
*Sources*: CAICYT (2008) and MCTIP (2008).

submitted to scientific meetings increased significantly during the same period. This suggests a positive effect of the demand-driven funds granted by the research authority that started to operate in 1993.

## Concluding Remarks

This chapter has outlined the complex Argentine public university management system. The reforms introduced since 1918 have contributed guaranteeing equality of access to higher education and pioneered, since then, reforms throughout Latin American universities. However, this complex system governing mass higher education, after nearly a century of sustained institutional expansion, has become too rigid to dynamically adjust to the recurrent academic quality demands of globalization. This chapter provides some elements to support the hypothesis that the current university management system is contributing, among other possible causes, to slow decision-making and the failure to adjust to the changing demands of globalization. The recent change in academic evaluation and research fund allocation policies illustrates some strategies that have been implemented by the federal government to encourage HEIs to comply with international quality standards. The availability of organized information on scientific production and the implementation by the universities of adequate performance indicators would be invaluable to assess the academic impact of these policies.

# Acknowledgments

This work was supported by the SECYT-PICT 2007, project number 803 (Argentina).

# References

Barro, Robert, and Xavier Sala-i-Martin. 2005. *Economic Growth*. Cambridge, MA: Massachusetts Institute of Technology Press.

Bernasconi, Andrés. 2008. "Is There a Latin American Model of the University?" *Comparative Education Review* 52 (1): 27–52.

Centro Argentino de Información Científica y Tecnológica. 2008. *Indicadores Bibliométricos elaborados por el CAICYT*. Buenos Aires: Centro Argentino de Información Cientifica y Tecnológica. http://www.caicyt.gov.ar.

Clark, Burton. 1983. *The Higher Education System: Academic Organization in Cross-National Perspective*. Berkeley: University of California Press.

Del Mazo, Gabriel, ed. 1957. *La Reforma Universitaria y la Universidad Latinoamericana*. Resistencia, Brazil: Universidad Nacional del Nordeste.

Delfino, José, and Héctor Gertel. 1996. "Modelos para la asignación del presupuesto estatal entre las universidades nacionales." In *Nuevas Direcciones en el Financiamiento de la Educación Superior*, ed. J. Delfino and H. Gertel. Buenos Aires: Ministerio de Cultura y Educación.

Fanelli, Ana. 2007. "Argentina." In *International Handbook of Higher Education*, ed. J. F. Forest and P. G. Altbach. New York: Springer.

Gertel, Héctor, and Alejandro Jacobo. 2007a. "Bases y Puntos de Partida Para la Calidad de la Gestión Universitaria en la Argentina." Unpublished manuscript, Department of Economics, Universidad Nacional de Córdoba.

———. 2007b. "Toward the Quality in the Management of Higher Education in Latin America: What Indicators Should Prevail?" *Revista Brasileira de Política e Administração da Educação* 20 (2): 29–42.

Halperin Donghi, Tulio. 1962. *Historia de la Universidad de Buenos Aires*. Buenos Aires: Eudeba.

Hicks, Diana M. 2007. "Global Research Competition Affects Measured US Academic Output." In *Science and the University*, ed. P. E. Stephan and R. G. Ehrenberg. Madison: University of Wisconsin Press.

Holm-Nielsen, and Thomas Hansen. 2003. *Education and Skills in Argentina: Assessing Argentina's Stock of Human Capital*. Washington, DC: World Bank.

Jones, Eric L. 1998. *Growth Recurring. Economic Change in World History*. Oxford: Oxford University Press.

Jones, Marc T., and Peter Fleming. 2003. "Unpacking Complexity Through Critical Stakeholder Analysis: The Case of Globalization." *Business and Society* 42 (4): 430–454.

Lewis, W. Arthur. 1978. *Growth and Fluctuations 1870–1913*. London: George Allen and Unwin.

Luftman, Jerry N. 1996. *Competing in the Information Age: Practical Applications of the Strategic Alignment Model*. New York: Oxford University Press.

Middaugh, Michael. 2001. *Understanding Faculty Productivity. Standards and Benchmarks for Colleges and Universities*. San Francisco: Jossey Bass Company.

Ministerio de Ciencia, Tecnología e Innovación Productiva. 2008. *Indicadores de Ciencia y Tecnología. Argentina 2007*. Buenos Aires: Ministerio de Ciencia, Tecnología e Innovación Productiva. http://www.mincyt.gov.ar.

Ministerio de Educación de la Nación. 1995. *Ley de Educación Superior 24.521*. Buenos Aires: Ministerio de Educación de la Nación.

———. 2008. *Indicadores Científico-tecnológicos de Universidades Nacionales 1998–2006*. Buenos Aires: Ministerio de Educación de la Nación.

Pérez Lindo, Augusto. 1998. *Políticas del Conocimiento, Educación Superior y Desarrollo*. Buenos Aires: Biblos.

Piffano, Horacio. 2007. *El Encuadre Normativo Económico Financiero de las Universidades Nacionales Durante los 50 Años de Vida de la AAEP*. Buenos Aires: Asosación de Argentina Economía Política. http://www.aaep.org.ar.

Pirani, Judith A., and Gail Salaway. 2004. "Executive Summary." Vol. 3 of *IT Alignment in Higher Education*. Boulder, CO: EDUCAUSE Center for Applied Research.

Schugurensky, Daniel. 2002. *1918, Students Ignite Democratic University Reform in Cordoba, Argentina*. Toronto: University of Toronto. http://www.oise.utoronto.ca.

Strehl, Franz, Sabine Reisinger, and Michael Kalatschan. 2007. "Funding Systems and their Effects on Higher Education Systems." OECD Education Working Papers 6. Paris: Organisation for Economic Co-operation and Development.

Tsai, Yau, and Sue Beverton. 2007. "Top-down Management: An Effective Tool in Higher Education?" *The International Journal of Educational Management* 21 (1): 6–16.

Waisman, Carlos H. 1987. *Reversal of Development in Argentina: Postwar Counterrevolutionary Policies and Their Structural Consequences*. New Jersey: Princeton University Press.

Waismann, Abraham. 1966. "Las Universidades Argentinas: Discurso a una Joven Nación." Córdoba: CICERO Artes Gráficas.

Weil, David. 2005. *Economic Growth*. 2nd ed. New York: Pearson Education.

Weinberg, Gregorio. 1995. *Modelos Educativos en la Historia de América Latina*. Buenos Aires: AZ Editores.

Williams, Gareth. 1990. *Le Financement de L'enseignement Supérieur*. Paris: Les Éditions de L'OCDE.

# Chapter 7

# The Tension between Profit and Quality: Private Higher Education in Oman

*Hana Ameen, David W. Chapman, and Thuwayba Al-Barwani*

## Introduction[1]

Since opening its first university in 1986, the Sultanate of Oman has invested in one of the fastest growing higher education systems in the world. By 2006 the country had 52 colleges and universities, enrolling 42,741 students (OMOHE 2007, 14; Sultan Qaboos University 2007, 3). This dramatic growth was financed mainly by national oil revenues, which in 2007 accounted for 67 percent of the national budget (OMOF 2007). However, 2004 projections were that these oil reserves will be largely depleted within the next ten years (OCHE 2004). New extraction technologies and recent discovery of some new oil reserves have extended this horizon, but the country's oil reserves are finite and production is already on the decline (Chapman et al. 2009).

Faced with the loss of oil revenues, Oman is seeking new strategies for sustaining its economy. To that end, Oman is investing heavily in higher education in the belief that it could provide highly educated workers to other countries. Omani leaders are counting on preparing an educated workforce at a quality level that will allow workers to successfully compete for jobs outside of the country.

Oman's substantial investment in higher education is motivated, in large part, by the government's belief that developing an alternative economy and remaining competitive in the international arena will depend heavily on a highly trained citizenry. The Omani government's goal is that by 2020, at least 50 percent of the 18–24 year old age group will continue to post secondary education, up from 19 percent (OCHE 2004). This is an ambitious goal, since 38 percent of the population is below the age of 15 (OMONE 2007).

Meeting the costs of these rising enrollments poses a challenge. Since its inception, public higher education in Oman has been free to students, and the government needed a strategy to slow and contain higher education costs while still encouraging system growth. In 1995, the government responded by legalizing private higher education and aggressively promoting its expansion. While private colleges and universities in Oman are for-profit institutions, each of which has an owner and shareholders, the government has provided generous incentives to encourage their growth. The government provides the land on which to build private higher education institutions (HEIs), exemption for such institutions from taxes for five years, and a capital grant to each one of 20 million Omani riyals (US$52 million). It also provides those institutions with scholarships that can be awarded to students from low-income families (Chapman et al. 2009). In return, these private institutions offer two advantages to the Omani government: shareholder investments contribute to the initial cost and operating expenses of the institutions, and, unlike their public counterparts, they are allowed to charge tuition.

While the growth of private higher education in Oman has been a major national achievement, there is concern about the quality of the education and the availability of career opportunities for the graduates (OCHE 2004). Omani HEIs already produce more college graduates annually than there are new and replacement jobs available, an oversupply projected to worsen as college participation rates increase (Al-Barwani et al. forthcoming). The questions now facing government and higher education leaders are: Where will these graduates find jobs? Are graduates of private HEIs well positioned to compete for the jobs that are available? How does the quality of private higher education compare to that of public higher education?

While investing in private higher education is an important part of Oman's strategy to prepare to transition to a post-oil economy and remain competitive in the world marketplace, little systematic study has been undertaken of the extent to which either senior leadership at the national level or academic staff at the institutional level regard private higher education as successful in providing Omani youth with a quality education that will allow them to compete effectively in the labor market. To that end, this study investigated the extent to which educators, government officials, and civic leaders believe that private higher education is adequately preparing Omani students for a less oil-intensive future.

The study is grounded in the multiple streams model of the policy formulation process (Kingdon 2003). Kingdon posits that public policy emerges from the intersection of three streams: (a) the *problem stream,* for example, the recognition that a problem exists; (b) the *solution stream,* for example, the formation and refining of policy proposals as potential solutions to the problem; and (c) the *political stream,* for example, the emergence of a consensus among various political forces that it is feasible to address the perceived problem with a particular solution. Public policy, in Kingdon's view, emerges when policy makers agree on the nature of the problem they are trying to solve, that it needs attention at a policy level, that there are potential solutions to the problem, particularly on the feasibility and efficacy of those alternatives, and that it is in the best interest of themselves and their country to address the problem.

Policy makers come to these views, in part, on the basis of what their colleagues and constituencies tell them are the problems needing policy-level attention and the judgments of these groups about the feasibility and acceptability of possible solutions. An objective of this research was to assess the extent to which there is congruence in the views of key constituencies regarding the quality of private higher education and in their judgments regarding the acceptability of possible solutions, if groups are found to hold different views.

## Methodology

The study employed a mixed-method design combining semi-structured interviews with a purposeful sample of 46 senior level educators, government officials, and private sector employers; surveying 252 college and university instructors; and document review.[2] The senior level respondents are influential in the formulation and implementation of higher education policy. The instructor group is largely the gatekeeper of the type and quality of instruction delivered in Oman college and university classrooms. Interviewees were selected on the basis of holding or having recently held a prominent leadership position in a government agency or private sector enterprise that had significant involvement with HEIs. The interview protocol collected respondents' views regarding the quality of higher education and strategy for quality improvement, academic and career advising of students, and "big" issues facing higher education in Oman. Interviews were tape-recorded and concurrent notes were taken.

In conducting the survey, the seven technical colleges in Oman were excluded from the study due to lack of relevance. Thirty-two of the remaining 46 colleges and universities (except Sultan Qaboos University [SQU]) were contacted by a representative of the Ministry of Higher Education and asked to have approximately ten faculty members and administrators complete the Oman Higher Education Access Questionnaire (OHEAQ) developed by the authors. The remaining 14 were excluded because it was logistically not possible to distribute and collect questionnaires in the time available due to their distant and/or remote locations. At SQU a larger sample was asked to complete the OHEAQ since it is the largest and only public university in the country. Across all institutions, HEI administrators who were in the interview sample did not complete a questionnaire, to avoid double counting. The OHEAQ contained 47 items and allowed us to collect data in four general areas: costs and financing higher education, quality of higher education, the need for flexibility and change in higher education, and issues in secondary to postsecondary transition. Data for 43 of the items were collected on a four-point Likert-type scale, while the remaining items were collected on two- or three-point scales.

The document review drew on Omani government documents, including a substantial number of government reports and internal publications that are difficult to access outside the country, published research about education in Oman and the Middle East, and the wider international research literature.

In summarizing interview data, content analysis was used to summarize the nature and intensity of respondent comments on the interviews. Frequencies and cross-tabulations were used to examine the questionnaire data. Chi square analysis was used to test the significance of differences among educators in private HEIs, public HEIs, and at SQU on key items. Findings from the interview and questionnaire are integrated into the discussion below.

# Findings and Discussion

When asked what they regarded as the biggest problem facing higher education in Oman, interviewees expressed a range of concerns but the most widely expressed apprehension was that the quality of Omani higher education is low (table 7.1). At the same time, a substantial number of interviewees recognized that Oman needed to find a way to absorb yet more secondary school graduates into higher education. The tension evidenced in table 7.1, for example—wanting to increase access while also needing

**Table 7.1** Content Analysis of Interview Responses: Concerns Facing Higher Education in Oman

| Quality | Number |
| --- | --- |
| Improve the quality of higher education; improving quality of private HEI | 19 |
| — Faculty evaluation and pay for performance | 1 |
| — Attracting and recruiting good professors | 1 |
| **Access** | |
| Need to extend access to higher education by absorbing secondary school graduates | 16 |
| — Students having to leave country to find higher education elsewhere | 1 |
| — Attracting international students to come to Oman | 1 |
| **External Efficiency—Employment** | |
| Align higher education with labor market needs | 8 |
| — Better performance of graduates | 2 |
| — Making graduates more competitive in the marketplace (especially abroad) | 2 |
| — Jobs for graduates of the HEI; shortage of jobs for graduates | 2 |
| — Improve transition to the workplace | 1 |
| — Attract foreign companies to Oman | 1 |
| — Gender discrimination in hiring | 1 |
| — Need better mid-career upgrading programs | 1 |
| **Funding for Higher Education** | |
| Funding higher education; loss of oil revenues to fund higher education | 6 |
| Challenge of diversifying to a post oil economy | 2 |
| Fundraising for private colleges and universities | 1 |
| Implementing cost share | 1 |
| Funding of students to attend higher education | 1 |
| Low level of salaries in HEIs | 1 |
| **Other HE Program and Management Issues** | |
| Using technology in education | 3 |
| Need for research capacity; lack of research at HEIs | 2 |
| Need programs for gifted children | 1 |
| Moving colleges to be able to offer MA and Ph.D. | 1 |
| Competing with the private sector in hiring top quality faculty | 1 |
| Greater flexibility in how HEI prepare students | 1 |
| HEI need to offer more diversified set of programs | 1 |
| Transferability of credit | 1 |
| What to do with faculty who do not make progress in their careers after being hired | 1 |

Continued

**Table 7.1** Continued

| Quality | Number |
| --- | --- |
| **Improvements in Secondary Education** | |
| Need better secondary school guidance programs | 1 |
| Better alignment of secondary preparation with higher education courses | 1 |
| Concern about success of basic education reform | 1 |
| Low quality of English and math skills of incoming students | 1 |
| More efficient use of resources in basic education | 1 |
| **Other Issues** | |
| Concern about overbuilding higher education system given demographics—could be overcapacity later | 1 |
| Credentialism in government hiring | 1 |
| Overly bureaucratic processes of MOHE | 1 |
| Clarification of rules on how a college can become a University | 1 |
| How to sustain culture of change when students who study abroad return to Oman | 1 |
| Availability of water | 1 |
| Recapturing benefits for Oman when graduates go to the Gulf and elsewhere to work | 1 |
| Developing international visibility for HEI | 1 |
| The diminishing value of higher education as more students achieve higher education | 1 |

to raise quality—is a classic dilemma in higher education. Quality can be raised by being more selective in the academic abilities of students who are admitted, therein reducing access. Conversely, extending access typically leads to the admission of more students with less robust academic abilities, resulting in downward pressure on quality. Trying to do both simultaneously poses serious challenges.

In a separate question, interviewees' views about quality were probed more deeply to determine whether the perceived problem of low quality was centered in public or private institutions. As table 7.2 indicates, responses were mixed. There was no consensus, in part because many interviewees thought quality varied widely from institution to institution within both sub-sectors. Differences among private institutions were seen as greater than the differences between public and private institutions. Among senior level educators, government officials, and private sector employers, the number who thought private was as good or better than public HEIs (N=5), who thought the opposite (N=4), and who did not think they could make a judgment (N=4) was about equal. There was considerable agreement,

**Table 7.2**   Content Analysis of Interviews with Senior Omani Leaders: Quality of Higher Education

| Interviewees' Assessment of Higher Education Quality | Number |
|---|---|
| Interviewees who believed quality of higher education is low | 17 |
| Of those, number who thought quality varied widely across institutions | 9 |
| Number who said private is as good as or better than public higher education | 5 |
| Number who regarded quality of public as better than private higher education | 4 |
| Number who said they were unable to judge the quality of higher education | 4 |
| **Suggested Solutions** | |
| MOHE should play a stronger role (MOHE needs to set clear standards for HEIs, raise standards, play a stronger role in quality assurance) | 11 |
| Make better use of international affiliations | 2 |
| Increase competition among private HEIs | 1 |
| Give scholarships that students can take to public or private institutions | 1 |
| Award incentive funds to encourage faculty to do research | 1 |
| Evaluation of HEIs should be outcomes-based | 1 |
| Colleges should be ranked; results should be published | 1 |
| Better staffing of Accreditation Council | 1 |

*Note*: This table is drawn from a larger table of results previously reported in Chapman et al. 2008.

however, that doing something to raise quality was the responsibility of the government, specifically the Ministry of Higher Education.

This concern about higher education quality was also reflected in the views expressed by college and university educators on the questionnaire (table 7.3). However, these educators held a much more unified view of the differences in quality between public and private institutions and a much more negative view of the quality of the private HEIs. Of the respondents, 64 percent did not consider the quality of most private colleges and universities to be comparable to the quality of public universities. Moreover, 66 percent thought that private colleges and universities have the least sufficient procedures for ensuring quality.

One of the most direct solutions would be more money, which could help institutions raise the quality of their instruction if funds were invested in stronger instructors, more student advising, and better laboratories and libraries. However, as discussed earlier, there is a tension between quality

**Table 7.3** Educators' Views on Quality in Omani Higher Education

| Question | Do not agree | Somewhat agree | Agree | Strongly agree |
|---|---|---|---|---|
| The quality of most private colleges and universities is comparable to the quality of public universities. (N=225) (Q 28) | 30.2 | 33.8 | 29.8 | 6.2 |
| To what extent do you think sufficient procedures are in place to ensure quality:[a] | Not at all | Somewhat | Very | Extremely |
| —At SQU? | 3.0 | 35.1 | 52.5 | 9.4 |
| —At other public colleges and universities? | 8.7 | 54.6 | 34.3 | 2.4 |
| —At private colleges and universities? | 10.7 | 56.1 | 29.4 | 3.7 |
|  | Do not agree | Somewhat agree | Agree | Strongly agree |
| It would be easy for the government to develop a reasonable estimate of a student's financial need. (Q 4) | 4.2 | 15.2 | 52.7 | 27.8 |

*Note*: a. Information from this table was previously reported in Al-Barwani et al. 2008.

and profit: the desire to invest in quality competes with the need to pay shareholders. While private HEIs can charge tuition, private higher education is expensive and tuition levels are effectively capped by the ability of many students to pay. Data from the document review suggests that raising tuition is problematic. Tuition can often be about US$5,000–$7,000 per year (OMOHE 2007); students going to private universities could be expected to pay up to US$780 per month. By way of comparison, per capita Gross Domestic Product is US$14,400 (CIA 2007). The beginning salary of a recent graduate with a bachelor's degree is about RO 600 (US$1,560) per month. Families in Oman generally have multiple children, with eight to ten not uncommon. Consequently, many families could be faced with having several children in college at once. At US$780 per month, with multiple children in college, private higher education is largely unaffordable for many families.

Students still might be able to cope with high tuition if there was a way they could borrow against future earnings to pay current college costs.

Indeed, the country's 2004 Higher Education Strategy Report pointed out the need to develop a reasonable way to estimate financial need, and had recommended that a loan program be developed (OCHE 2004; Preddy 2004). That recommendation was removed from the final version of the report in response to skepticism regarding the practical issues of managing a loan program. However, over the last four years, these views have perhaps been shifting. Chapman et al. (2009) report that senior government, education, and business leaders now express increasing support for greater cost sharing on the part of families and students, though there is still concern over practical aspects of fairly assessing family financial need and securing repayment. Results of the present study indicate that college and university instructors shared the view of senior leaders. They were supportive of students paying more of the cost of their education, and of the feasibility of loans to assist students with insufficient funds. Moreover, a majority of those educators did not share the skepticism about estimating students' financial need. Most (80.5 percent) thought such an estimate could be developed.

Findings from this study indicate that there is wide agreement among senior level educators, government officials, and private sector employers that quality of Omani higher education, overall, needs to be raised. However, there is little agreement within this group as to the relative quality of public and private institutions of higher education. Instructional staff within Oman's colleges and universities shared the concern that, across all institutions, higher education quality was low, but expressed considerably greater agreement that the main problem centered in the private HEIs.

The government has been mindful of the need for quality since the time private higher education first was introduced. However, two decisions made at the time the private sector was legalized have had unanticipated consequences that negatively impact current government efforts to promote quality. First, all private HEIs were established as for-profit enterprises. On one hand, this made sense. A principal reason to establish a system of not-for-profit private HEIs (such as in the United States) is to encourage private philanthropy in support of higher education. Donors receive tax breaks for their gifts. However, Oman does not have an individual income tax and corporate taxes are fixed at 12 percent, low by Western standards. Consequently, there is little need for tax advantages, and little reason to establish a special not-for-profit designation.

On the other hand, establishing private HEIs as for-profit enterprises has created a potential conflict between the desire of stockholders in private HEIs to make a profit, and the desire of the private institutions to raise quality. A notable number of interviewees (N=8) expressed concern that the profit motive threatened the quality of private HEIs. From their perspective, private colleges face a tension between investing in quality instructors, facilities, and programs and providing investors with a reasonable return on their investment.

The second decision that carried unanticipated consequences was to require all private HEIs to have an affiliation with an international counterpart, an arrangement that was intended to serve as a quality assurance mechanism. While the specific arrangement varied by institution, the general design was that these international affiliates either directly developed, or, in some cases, reviewed the overall curriculum and individual course content, and would sometimes be involved in the appointment of administrative and teaching staff. This involvement was intended as a way of assuring that programs and instruction in the Omani institution were comparable to the affiliated institution.

In the absence of a national accrediting mechanism, this affiliation system provided a system of quality assurance. In return, the international affiliate was paid for their services, either as a fixed fee, a percent of profits, or on a per capita enrollment basis. At times, these fees were substantial. Some affiliates received over US$100,000 per year for their oversight services. In some institutions this affiliation system worked well. The affiliate was attentive, involved, and committed to quality assurance and improvement. Other affiliates, however, took a more relaxed approach to the arrangement, and quality assurance was weak.

The tension between investing in quality versus rewarding shareholders in combination with an uneven quality assurance mechanism is further complicated by the tendency of private HEIs to attract lower quality students than their public counterparts. The Omani government continues to pay the costs of students attending public HEIs while most students in private HEIs pay tuition. Consequently, attending SQU (the only public university) or a public college is more attractive. As a result, public HEIs attract a strong applicant pool and are able to select the strongest students in terms of academics. Those not selected for government sponsored public higher education must either attend a private college or study abroad. Working with academically weaker students raises the cost of instruction for private HEIs, as these students need more remediation, extra tutoring, and more intensive academic support.

The personal calculus of whether the investment in higher education is worthwhile is closely tied to families' expectations about the employment their student will be able to find upon graduation. Do the long-term benefits of a college education offset such a substantial up-front cost? The results reported above help identify a further dilemma faced by those enrolling in private higher education. Specifically, there is already an oversupply of bachelors-level graduates relative to the employment opportunities at that level in Oman, a situation projected to worsen. By some estimates, if the college participation rate rises to 30 percent of those students now in Grade 10 (well within government targets), there could be over 5,000 more graduates than the number of available public or private sector jobs in the year they would graduate from college, even assuming aggressive efforts to replace expatriate workers with Omanis are successful (Al-Barwani et al. forthcoming).

The expectation of the government is that those unable to find a job in Oman would emigrate to find employment in the Gulf Region or beyond. Interview results indicated that this is widely understood by the senior government, education and civic leaders. Given the projected over-production of college graduates relative to the domestic labor market, employment-oriented emigration is an essential part of the longer-term strategy for adjusting to the decline in oil production. It is likely that the need to emigrate to find employment will fall more on private than on public college and university graduates, for three reasons. First, graduates of SQU and other public HEIs have an advantage in competing for employment within Oman because they are widely viewed as having a higher quality education. Second, there has been a history of employers favoring public HEI graduates in public sector hiring, largely because they recognize that the more academically capable students still seek admission in public institutions where the government pays costs. Finally, women tend to out-perform men in secondary school and, consequently, tend to out-number men in public higher education. This leaves relatively more men seeking to enroll in and graduate from private HEIs. For cultural reasons men are more likely than women to leave the country to seek employment.

It should be noted that the presumption that graduates will emigrate for employment has not been fully tested. To date, emigration has been modest. According to government records, only 314 Omanis left for employment opportunities in Qatar and Dubai in 2006 (through government sponsored programs), even though there were over 17,000 job seekers unable to find jobs in the country (OMOM 2007; Al-Barwani et al. forthcoming). Nonetheless, assuming emigration increases as the oversupply of graduates grows, those leaving the country will be competing to obtain and hold employment within an international labor market.

Success in finding and holding employment in the international setting will be based on perceived quality of education, job performance, and work ethic. Omani graduates will not have the benefits of the national "Omanization" policy (encouraging the replacement of expatriate workers with Omanis) or the strong employment protections afforded Omani citizens working in Oman. Job retention based on performance may be a problem. The 2005 Arab World Competitiveness Report suggests that educational preparation and a positive work ethic have not been strengths of Omani workers (Lopez-Claros and Schwab 2005). Omani business leaders (e.g., the respondents in the Arab World Competitiveness Report) regarded the Omani workers as having inadequate educational preparation and a poor work ethic. Consequently, the international competitiveness of Omani workers is at risk. If Omani graduates are unable to secure and hold employment abroad, Oman's larger strategy of investing in higher education as a way to build an alternative economy that will be competitive in a post-oil era is put at risk.

Private HEI graduates are the most likely to leave Oman to find employment, and as noted above, their success will depend, to a large extent, on the quality of their educational preparation. Despite the widespread recognition that the quality of private higher education in Oman is low; however, there are organizational and structural impediments to raising quality. Among them are the potential conflict of interest of policy makers who also hold a financial interest in private HEIs, skepticism about the willingness of private colleges to forgo profit to invest in higher instructional quality, conflicts between the past and presumably future quality assurance systems, and a reluctance among some government agencies to adopt student loan schemes that might give private colleges more flexibility in setting their tuition levels.

## Conclusions and Recommendations

The web of challenges now faced by HEIs in Oman is generally well understood by both senior national leaders and institution-level college and university educators. Low quality of private higher education and the need for additional funding are widely viewed as the dominant challenges now facing the system. Viewed through the lens of Kingdon's model, the findings of this study suggest that there is more consensus within the problem stream than within the solution stream. Strategies for raising quality and increasing funding tend to have a web of consequences such that actions needed to resolve any particular problem are likely to be unpopular with key constituencies. Their resistance raises the political costs of corrective action, making it difficult to develop a consensus on the solution stream. Already, however, it is clear that some actions to promote quality should at least be considered.

First, there is widespread agreement that Oman needs a more effective system for quality assurance in higher education. To make effective progress in that direction, it may be necessary to reduce the potential for conflict of interests among education leaders who are also shareholders. This might involve limiting government officials who have any level of oversight of higher education (public or private) from being a shareholder or owner in any private college or university. Alternatively, the government might require officials to make a full public disclosure of any direct or indirect financial interest in any HEI. This could reduce self-interest of decision makers who must find the balance between profit and quality.

Second, it is not always clear whether the problems graduates encounter in finding employment are due to the quality of their higher education or their inexperience in the international job market. As Oman moves toward an era of increased employment-oriented emigration, it will be important that both the government and HEIs implement stronger college and career advising at both the secondary and postsecondary levels. Students entering college need to more clearly understand their domestic and international employment prospects, the educational qualifications required for entry to those jobs, and the career ladders that characterize careers in those fields. If better career advising of graduates leads to better employment success, the potential backpressure on private HEIs may be reduced.

Third, HEIs can do much to help students understand the growing importance of merit and performance in job success. Depending on the institution, this might be accomplished by implementing stricter performance standards for course and program completion. The standards used to judge student performance in college should model what is likely to be their subsequent experience in the workplace. This is perhaps the most direct tension between profit and cost, as students unable to meet higher performance standards may withdraw. HEIs need to assess the cost-benefits of keeping weak students enrolled versus raising the percent of successful graduates.

Finally, more money is sometimes, though not always, an important input in raising instructional quality. When additional funds are part of the solution, private HEIs need additional means of raising those funds. To that end, the Omani government needs to revisit student loans as an option available for students lacking sufficient funds to attend private higher education.

# Notes

1.  The authors express appreciation to Ruqaia Hilal Alma'ani (Oman's Ministry of Higher Education) for her assistance in data collection and to Jessica Werner (University of Minnesota) for her assistance in data analysis.

2. The methodology employed in this study and described here has previously been presented in Al-Barwani et al. (forthcoming).

# References

Al-Barwani, Thuwayba, David W. Chapman, and Hana Ameen (forthcoming). "Strategic Brain Drain: Implications for Higher Education in Oman." *Higher Education Policy.*

Central Intelligence Agency. 2007. *The World Factbook.* Washington DC: Central Intelligence Agency. https://www.cia.gov.

Chapman, David W., Thuwayba Al-Barwani, and Hana Ameen. 2009. "Expanding Post-Secondary Access in Oman, and Paying for It." In *Financing Access and Equity in Higher Education,* ed. J. Knight. Rotterdam: Sense Publishers.

Kingdon, John W. 2003. *Agendas, Alternatives, and Public Policies.* 2nd ed. New York: Longman.

Lopez-Claros, Augusto, and Klaus Schwab. 2005. *The Arab World Competitiveness Report 2005.* Geneva: World Economic Forum and New York: Palgrave Macmillan. Cited in *World Ministry of Finance 2007. The State's General Budget for the Fiscal Year 2007.* Muscat: Sultanate of Oman.

Oman's Council on Higher Education. 2004. *The Strategy for Education in the Sultanate of Oman, 2006–2020.* Muscat: Sultanate of Oman.

Oman's Ministry of Finance. 2007. *The State's General Budget for the Fiscal Year 2007.* Muscat: Sultanate of Oman.

Oman's Ministry of Higher Education. 2007. *Higher Education Institutions in the Sultanate of Oman, 2007.* Directorate General of Private Universities and Colleges. Muscat: Sultanate of Oman.

Oman's Ministry of Manpower. 2007. *Annual Report, 2006.* Muscat: Sultanate of Oman.

Oman's Ministry of National Economy. 2007. *Monthly Statistical Bulletin.* Muscat: Sultanate of Oman.

Preddy, George. 2004. "Funding Of Education in the Sultanate of Oman: The Context, Issues, Tasks, and Outcomes." Background paper prepared for the Higher Education Strategy Development, Council on Higher Education, Sultanate of Oman, Muscat.

Sultan Qaboos University. 2007. *Statistical Yearbook, 2006.* Muscat: Sultanate of Oman.

# Chapter 8

# Surviving Austerity: Kenya's Public Universities and the Competition for Financial Resources

*Gerald Wangenge-Ouma*

## Introduction

Several aspects and trends characterize the global context of higher education. These include, among others, an increase in the number and types of higher education providers and consumers, marketization and privatization, a worsening global economy, and declining state financial support. Declining state financial support is perhaps the most critical aspect of the present constitutive environment of public universities, mainly because it constitutes a threat to these institutions' commitment to their core missions. Declining state financial support for public universities therefore calls for both strategic and adaptive responses by institutions to guarantee their well being in a competitive environment.

This chapter focuses primarily on how public universities in Kenya compete for resources, especially financial resources, in response to declining state funding. The author discusses the main strategy through which Kenya's public universities compete for financial resources—that is, the dual track tuition fee programs in which privately sponsored, full fee-paying students are admitted alongside state subsidized students. Also discussed are the two main ways in which public universities seek competitive advantage in implementing the dual track tuition fee programs, specifically: (a) the establishment of franchise arrangements with private non-university institutions and expansion via the establishment of satellite campuses, and (b) the introduction of "demand driven" programs. The author concludes by discussing how the pursuit of economic self-determination through dual track tuition fee programs has occasioned the phenomenon of goal displacement, especially with regard to quality.

# Higher Education Funding in Kenya

Declining state funding of public higher education is a global reality (Slaughter and Leslie 1997; Clark 1998; Santiago et al. 2008). This trend is characterized by two main factors: (a) declining state funding and, (b) increasing or steady allocations not matching the needs of the rising expenditures of the universities.

The degree of decline in state funding for public higher education varies from country to country. In Uganda, an average of 10 percent of the total education budget went to the higher education subsector between 1998 and 2003, compared to the early 1990s when this amount was 19 percent (Carroll 2006). In South Africa, as a percentage of Gross Domestic Product (GDP), state funding of higher education declined from 0.82 percent in 1996 to 0.67 percent in 2006 (Wangenge-Ouma and Cloete 2008). Paulo Santiago et al. (2008) report that between 1995 and 2004, public expenditure per higher education student in Chile, Hungary, Austria, and the United Kingdom declined by 34, 28, 27, and 19 percent respectively.

Kenya's public universities have not been spared from declining state funding. From 1996 to 2000, state funding for public higher education as a percentage of GDP averaged 0.94 percent and reduced to 0.74 percent in the period from 2001 to 2005 (Wangenge-Ouma 2008a). The Kenyan government's incapacity to meet the financial needs of public universities is exemplified by its inability to meet the universities' budgetary requests, as epitomized by the case of Jomo Kenyatta University of Agriculture and Technology (JKUAT) in which the government failed to meet the university's requests by a large margin in five out of six cases between 2000 and 2005 (Ouma, 2007). The phenomenon of significant budget shortfalls is experienced by most of Kenya's public universities (Ouma 2007). Assuming that the universities' budgetary requests are not exaggerated, this phenomenon is perhaps the surest indication that the state cannot be relied upon to meet the financial requirements of public universities.

Due in part to the government's insufficient allocations to public universities, many Kenyan public universities were in a state of financial emergency from the 1990s through the mid 2000s. In both 1998 and 2001, the Auditor-General described JKUAT and Egerton University as being "technically insolvent" (Ouma 2007). As of 31 May 2005, the University of Nairobi was running a debt of approximately US$23,228,683 accumulated over the years. In the 2004–2005 financial year, Kenyatta University (KU) experienced a deficit of approximately US$2,857,143 (Wangenge-Ouma 2008b).

The second trend that characterizes the decline of state funding for public universities—increasing or steady allocations not matching the needs of rising expenditures at public universities—generally does not apply in the Kenyan case at present. This phenomenon occurs when state funding for higher education has not declined in real terms, but allocations are still inadequate given expansion and enrollment demands on the subsector, escalating costs of higher education that far outstrip inflation, and demands for research and services provided by higher education (Benjamin and Carroll 1998). This phenomenon is partly evidenced by a situation in which declining per-student expenditure persists even when the growth rate of public higher educational expenditure is positive (Varghese 2001).

# Global Pressures, Local Realities

Generally, declining state funding of higher education may be described as a result of a convergence of global pressures and local socioeconomic and political realities. Globally, the establishment of neoliberalism as the de facto economic logic has discredited the public model of financing higher education. Neoliberalism, which "advocates for privatization, marketization and performativity; and the shift of the cost of public services (for example, higher education) from the state onto the individual" (Wangenge-Ouma 2008b, 216), privileges the adaptation of public universities to a competitive market situation. In poor, dependent, and dominated economies, higher education's encounter with neoliberal globalization cannot be separated from the coercive influence of international financial institutions (IFIs), mainly the World Bank (WB) and the International Monetary Fund (IMF) (Aina et al. 2004; Wangenge-Ouma 2008b), which have been the primary purveyors of the neoliberal economic logic for a long time.

Akin Tade Aina et al. (2004) argue that the imposition of the neoliberal economic doctrine on Africa went hand-in-hand with many African countries ceding their responsibility for social and economic development management to IFIs. Policy formation thus became delocalized as this role was largely appropriated by the IFIs. The decision to diminish funding for public higher education is a case in point of a delocalized policy imperative.

Local realities have made it equally difficult for many countries to continue with large-scale funding of public higher education. Especially in Africa, sluggish economic growth, devaluation of currency exchange rates, deep external debts, intersectoral competition for public funds, and growth in the higher education subsector have made it impossible for countries to continue with substantial funding of higher education (Wangenge-Ouma 2008b).

Kenya offers a germane example of how local realities precluded the government's continued large-scale funding of public higher education. Beginning in the early 1980s, the economic conditions in Kenya were not favorable for the government's continued large-scale funding of public higher education. The period of 1980 through 1985 was one of low GDP growth, averaging about 2.5 percent per annum with some years experiencing negative growth rates. This was then followed by the period of structural adjustment programs (SAPs) from 1986 to 1989. SAPs entailed, in the main, a significant reduction of state expenditure on social services (including higher education), and privatization of state-owned industries. The 1990s were an economic nightmare for Kenya. During this period, the economy was characterized by sluggish growth and the country witnessed very high inflation rates (Republic of Kenya 2001, 2003). Confronted with these unfavorable economic conditions, the Kenyan government was unable to continue with extensive funding of higher education.

Through several policy papers published at this time, the Kenyan government exhorted public universities to seek more funding from market sources. For instance, the 1997–2010 Master Plan states that:

> Universities will be encouraged to develop non-public sources of their revenues, including income-generating activities (such as returns from research and consultancies with industry and employers, services to the community, agro-based production, manufacturing for the market, including making equipment for use in schools, hiring out university facilities); grants and donations from NGOs and well-wishers; and funding from alumni associations. (Republic of Kenya 1998, 110)

The government thus sought to change its role in higher education financing by devolving financial responsibilities to public universities. Higher education institutions (HEIs) were discouraged from relying solely on public funding and had to compete for resources from non-public sources.

In summary, global and local realities have conspired against continued large-scale state funding of public universities in Kenya, making seeking financial resources from the market an imperative, and leading to the rise of the "enterprise university" as an organizational archetype (Clark 1998; Ouma 2007).

## Competition for Resources

Competition for resources was not a path the universities opted to choose; instead, it is an undeniable feature of the new context of higher education briefly described above. Other than declining state funding, the new context of higher education is also marked by the presence of multiple and various higher education suppliers: public and private, for-profit and not-for-profit, distance and contact institutions, all of which need, and compete for, resources.

Kenya's higher education system is also characterized by an increase in both the number and variety of providers. With only one public university in 1970, the country now has seven public, and approximately 21 private, universities. Several foreign universities also ply their trade in Kenya. The landscape of higher education in Kenya is therefore quickly becoming a congested one—a condition that heightens competition for resources.

Competition for resources, as postulated by resource dependence theory (Pfeffer and Salancik 1978), is geared at guaranteeing organizational survival. Considering the conditions of declining state funding discussed in the preceding section, the emergent vulnerability demands that public universities mitigate the unreliable patronage by the state by competing for resources, especially from the market.

## Shifting Resource Dependence from the State to the Market

The main way through which Kenya's public universities have attempted to weaken their resource dependence on the public purse is through the implementation of dual track tuition fee programs. These programs entail the admission of privately sponsored, full fee-paying students over and above the quota of students that receive a government subsidy. The privately sponsored students are enrolled in programs generally known as "parallel programs."

The University of Nairobi (UoN) was the first institution to introduce parallel programs in 1998. Crispus Kiamba (2004) correctly locates the introduction of parallel programs against the background of the biting fiscal hardship UoN was experiencing, coupled with the government's declared inability to continue large-scale funding of the higher education subsector. The need to overcome this debilitating financial situation prompted the university to form a committee in 1994 to recommend ways and means of earning nongovernment revenue. Kiamba (2004) reports on the committee's attraction to the concept of the "entrepreneurial university," where emphasis is placed on the identification of a university's resources and their capacity for commercial development. The committee recommended establishing continuing education programs as a top priority area for implementation. The existence of untapped capacity in the university (evenings and weekends) provided the opportunity for the introduction of parallel programs. All of Kenya's seven public universities have since started parallel programs. It is reasonable to conclude that UoN served as a test case and legitimated the practice for other public universities to follow.

Aside from the existence of unused capacity in the universities, the introduction of parallel programs was also encouraged by the local existence of a broad, yet unmet, demand for higher education. On average, only about 26 percent of qualified candidates gain admission into Kenya's public universities with a government subsidy. In the 2001–2002 academic year, out of 40,496 qualified candidates, only 11,147 (27.5 percent) were admitted with a government subsidy. In the 2002–2003 academic year, out of 42,158 qualified candidates, only 10,966 (26 percent) were admitted, and in the 2003–2004 academic year, out of 42,721 qualified candidates, only 10,947 (25.6 percent) were admitted with a government subsidy (Wangenge-Ouma and Gravenir 2006). One trend is clear: whereas the number of qualified candidates is increasing, the percentage of candidates gaining admission with a government subsidy continues to decrease. The nearly 21 private universities in the country admit only approximately 20 percent of all qualified candidates. Therefore, as the parallel programs were introduced, there was a significant unmet demand for university education, which provided a ready market for the universities to exploit.

Since the introduction of parallel programs, Kenya's public universities have managed to generate significant funding and have decreased their dependence on state funding to various extents. Between 2001 and 2005, parallel programs accounted for an average of 27 percent of UoN's total revenue, 15.4 percent at Kenyatta University and 15 percent at JKUAT (Ouma 2007). Parallel programs have therefore changed many of Kenya's public universities from the previous unenviable situation of penury to some degree of self-reliance.

Considering the relatively high fees charged for these programs, and the competitive strategies employed by public universities to attract more students, motivated by the need for capital accumulation (see subsequent section below), it can be argued that parallel programs are a form of trade in higher education. Parallel programs thus align with neoliberalism's treatment of higher education as a private, tradable commodity to be purchased by a consumer, a product to be retailed by academic institutions (Altbach 2002), or, as Michael Peters and Peter Roberts (1999) put it, "something to be produced, packaged, sold, traded, outsourced, franchised, and consumed" (175).

It should be noted that dual track tuition fee policies are not only a Kenyan phenomenon. Similar programs are in vogue in several African countries—such as Uganda, Tanzania, Egypt, and Malawi—and in several other countries around the world, such as Australia, India, and Russia. It can therefore be argued that dual track tuition fee programs are gradually becoming a favored strategy by public universities to mitigate declining state funding.

## Seeking Competitive Advantage

As aforementioned, all of Kenya's public universities have turned to the student market to try to reduce their resource dependence on government funding. As a result, stiff competition has arisen for parallel students, since the number of parallel students a university is able to enroll determines the amount of revenue earned.

The public universities have implemented two main competitive strategies—collaborative alliances with other learning institutions, especially private non-university institutions, and expansion. Expansion as a strategy for competitive advantage takes place mainly through the establishment of campuses in strategic locations, and introduction of new study programs. The rationale for the two key strategies of collaborative alliances and expansion is to increase the universities' share of the student market and the consequent attainment of optimal revenue generation.

## Collaborative Alliances

Collaborative alliances are competitive strategies in which organizations exchange "resources" and attempt to cope with the turbulence of their environments (Pfeffer and Salancik 1978). From the perspective of resource dependence theory, collaboration is an exchange of resources, which occurs when the organizational decision-makers believe that joint involvement can protect, and perhaps enhance, key organizational resources (Connolly and James 2006). Since organizations are embedded in an environmental context composed of a variety of actors, and organizational survival is dependent on extracting resources from this environment, the actors become mutually dependent and work to ensure themselves a sufficient flow of resources (Pfeffer and Salancik 1978).

Several of Kenya's public universities have entered into resource-beneficial alliances with other educational institutions. The main form of these alliances involves franchise arrangements, especially with private non-university institutions. With more than 18 such ventures, JKUAT has the highest number of collaborative alliances, in which students enrolled in the franchised colleges are awarded certificates of the partner university. The courses offered are mainly in information and communication technology and business studies—the so-called demand-driven programs. In these collaborative alliances, the middle-level colleges—which are not certified by the country's accrediting body (the Commission for Higher Education) to offer degree courses—offer these courses in the name of JKUAT.

As is the case in other franchise arrangements, the relationship between JKUAT and its franchisees is mainly financial. The university receives a percentage of all the fees payable by students registered in courses for which the university's certificates are issued. The arrangement that JKUAT receives a percentage of *payable* fees, as opposed to fees that have actually been paid, cushions the university from problems of student debt at its franchises. It is therefore incumbent upon the franchises to collect all fees so as to meet their financial obligations to JKUAT. Other than collecting tuition fees, the franchises incur almost all the operational costs. Thus, in these collaborative alliances, universities are keen to minimize expenses and maximize revenue returns.

Consequently, through collaborative alliances, some of Kenya's public universities have been able to garner previously unavailable competitive advantages. The universities benefit from the widespread nature of these franchise colleges, and their strategic location in major towns—which allows them to attract many students. Some of these colleges offer courses that the universities do not provide on their campuses, meaning that the universities are able to benefit from these colleges' capacities. Through these interorganizational dependencies, universities are now able to obtain and retain revenue they would otherwise not obtain, and are able to do so efficiently. Overall, through engaging in interorganizational arrangements, Kenya's public universities have co-opted potential competitors into allies in the struggle for scarce resources, and have both gained access to and exploited the complimentary assets that the colleges bring to the arrangement.

## Expansion: New Campuses and Academic Programs

Through what may be described as "expansionism," Kenya's public universities have also sought to optimize tuition revenue through establishing branch and satellite campuses and creating new academic programs. Almost all the public universities have established satellite and branch campuses in strategic locations, especially in major cities such as Nairobi, Mombasa, Kisumu, Eldoret, and Nakuru, where there is high demand for university education, mainly by persons who are working.

The establishment of satellite campuses, especially in strategic locations, may be described as a business tactic of moving closer to potential clients. As a prime example of expansionism, Moi University has the highest number of satellite campuses, which are distributed throughout the country: Nairobi, Eldoret West, Kericho, South Nyanza, Mt. Kenya, and Kitale. Most of these campuses offer a limited range of "demand-driven" courses.

Public universities have also sought to expand their share of the student market by introducing new academic programs, usually in fields beyond their core areas of strength, such as health sciences, legal studies, information and communication technology, management and business studies (Wangenge-Ouma 2008a), and by implementing new modes of delivery, especially open and distance learning (as at Kenyatta and Egerton Universities).

An interesting phenomenon of expansion is the proliferation of bridging and preuniversity courses meant to assist students who would not otherwise qualify for university admission. Viewed from the perspective of income generation, the significant increase in bridging courses could be argued as a strategy by public universities to guarantee a constant flow of parallel students. Partly because of these strategies, and also because of the high demand for parallel programs, several of the public universities have since enrolled higher numbers of parallel students than the government-subsidized ones. From academic years 2001–2002 to 2005–2006, UoN had 25,517 parallel students as compared to 20,882 who were government-subsidized (University of Nairobi 2005). In 2005, government-subsidized students at Kenyatta University constituted about 44 percent of the student population. The rest come from various programs that charge full fees (Kenyatta University 2005).

# Goal Displacement

Not uncommonly, the survival strategies employed by universities occasion the phenomenon of goal displacement, whereby important institutional and policy goals—such as quality, equity of access, diversity, and transformation—are disregarded in pursuit of the "more pressing" problem of economic survival. This is in line with resource dependence theory's argument that despite other goals and aims, survival is the core objective of every organization (Pfeffer and Salancik 1978). The organization must first survive in order for it to achieve anything else.

There are various goals that may be "displaced" by the prioritization of economic survival by public universities, including the pursuit of policies such as access for marginalized communities, knowledge production, and quality. This section focuses on the "displacement" of the quality goal, given its significance toward achieving the universities' core mission as educational institutions.

Since the commencement of parallel programs, erosion of quality is probably the most critical and also the most frequently raised concern. A recent study by Gerald Wangenge-Ouma (2006) demonstrates that the main way in which parallel programs have impacted the quality of Kenya's higher education is the surge in student numbers and the proliferation of new academic programs in a context of limited human and institutional capacity, especially teaching staff and physical and pedagogical resources. The "capacity conundrum," cannot, however, be blamed entirely on parallel programs; capacity limitations of Kenya's public universities predate the introduction of such programs (Wangenge-Ouma 2006, 2008a). The universities did not have enough physical facilities, and most of those available were suffering decay following many years of neglect, arising mainly from under-funding. Equally important, the universities did not have enough teaching staff. Therefore, parallel programs were generally grafted onto a university system that was inimical to the attainment of quality teaching and research (Wangenge-Ouma 2006, 2008a).

As a result of attaching parallel programs onto a system with significant capacity challenges, the subsequent pressure seems to have made the quality situation worse. The universities have thus pursued economic self-determination at the expense of quality. The quality challenges that have been made acute by the introduction of parallel programs include, among other things, staff shortages and inadequate facilities.

In terms of staff shortages, Kenya's public universities have not been able to maintain a sufficient corpus of teaching staff, and all universities have many unfilled teaching positions. In the 2004–2005 and 2005–2006 academic years, UoN had a faculty establishment of 1,797 but with only 1,207 and 1,267 faculty in posts, respectively. In the same years, JKUAT had a staff establishment of 1,863 but with only 1,220 and 1,232 staff in posts, respectively (Wangenge-Ouma 2006). The staff shortages were acute at senior levels at both institutions. Staff shortages vis-à-vis the explosion in student numbers have had various results with implications for quality, such as lecturers having huge teaching loads and being asked to teach year round without a break, eventually suffering burn out; discontinuation of tutorials against a context of increasing number of weak students; and recruitment of unqualified people (Wangenge-Ouma 2006, 2008a).

Staff shortages are complicated by inadequate facilities. Even though several of Kenya's public universities have developed and acquired facilities in an attempt to accommodate the soaring number of students, facilities have generally failed to match the rate of increase in enrollments. The most strained facilities are classrooms. In addition, important facilities for science courses and library books are either lacking or inadequate. For science courses, the problem of facilities ranges from inadequacy to total unavailability. Science students are thus forced to crowd around the few sets of equipment available during experiments (Wangenge-Ouma 2006, 2008a). Generally, the challenge for many faculty is how to handle large classes with inadequate facilities and supplies, and for students, how to eke out a rewarding educational experience in such conditions.

By pursuing economic self-determination through parallel programs, in advance of well-developed capacity, the subsequent pressure seems to have inverted the universities' priorities from pursuing robust, high quality scholarship, to survival via capital acquisition and accumulation.

# Conclusion

Attempts to restore some equilibrium during conditions of austerity are an essential character of organizations, including public universities. For public HEIs, the generation of revenue from market sources constitutes an important measure for mitigating financial instability, arising mainly from declining state funding. As shown in the analysis presented here, Kenya's public universities have assumed market-like behavior as the preferred strategy for addressing waning state financial support. The market-like behavior of universities is both a strategic and adaptive response, which, as shown in this analysis, is largely driven by organizational survival. As a result of self-interested economic strategies that privilege organizational survival above all else, important institutional and policy goals, especially with regard to quality, seem to have been displaced.

Kenya's public universities are therefore facing a dilemma: how to survive austere conditions without compromising their expected role as citadels of excellence. Although the neoliberal economic logic has discouraged large-scale funding of higher education, in a context such as Kenya's, the need for improved state funding—especially for capacity building (infrastructural and staff development)—cannot be overemphasized. However, considering that substantial state funding for public universities may not be possible, especially in the present context of global economic decline, it remains incumbent on the public universities to institute measures that will mitigate the negative effects of their income earning initiatives. Such measures may include differentiation of focus and specialization of programs so that the universities concentrate on what they can do best without spreading themselves too thin. The nature of the ongoing marketization of Kenya's public higher education also brings into sharp focus the government's role in consumer protection and quality assurance. It is therefore important that the Kenyan government institutes some forms of regulation to help guard against the run-away marketization that seems to be taking place in the country's public universities without regard to quality.

# References

Aina, Akin Tade, Sethy L. C. Chachage, and Elizabeth Annan-Yao, eds. 2004. *Globalization and Social Policy in Africa*. Pretoria: UNISA Press.

Altbach, Philip G. 2002. "Knowledge and Education as International Commodities: The Collapse of the Common Good." *International Higher Education* 28 (Summer): 2–5.

Benjamin, Roger, and Steve Carroll. 1998. "The Implications of the Changed Environment for Governance in Higher Education." In *The Responsive University: Restructuring for High Performance*, ed. W. G. Tierney. Baltimore, MD: Johns Hopkins University Press.

Carroll, Bidemi. 2006. "Harnessing Private Monies to Fuel University Growth: A Case Study of Makerere University." Paper presented at DBSA/HSRC/Wits/NEPAD Conference: Investment Choices for Education in Africa, Johannesburg, September 19–21, 2006.

Clark, Burton R. 1998. *Creating Entrepreneurial Universities: Organisational Pathways of Transformations*. Oxford: Pergamon.

Connolly, Michael, and Chris James. 2006. "Collaboration for School Improvement. A Resource Dependency and Institutional Framework of Analysis." *Educational Management Administration and Leadership* 34 (1): 69–87.

Kenyatta University. 2005. "Kenyatta University Strategy and Vision Plan, 2005–2015." Nairobi: Kenyatta University.

Kiamba, Crispus. 2004. "Privately Sponsored Students and Other Income Generating Activities at the University of Nairobi." *Journal of Higher Education in Africa* 2 (2): 53–74.

Ouma, Gerald Wangenge. 2007. "Reducing Resource Dependence on Government Funding: The Case of Public Universities in Kenya and South Africa." PhD Diss., Graduate School of Education, University of Cape Town, Cape Town.

Peters, Michael, and Peter Roberts. 1999. *University Futures and the Politics of Reform in New Zealand.* Palmerston North: Dunmore Press.

Pfeffer, Jeffrey, and Gerald R. Salancik. 1978. *The External Control of Organizations: A Resource Dependence Perspective.* New York: Harper and Row.

Republic of Kenya. 1998. *Master Plan on Education and Training, 1997–2010.* Nairobi: Jomo Kenyatta Foundation.

———. 2001. *Totally Integrated and Quality Education and Training. Report of the Commission of Inquiry into the Education System of Kenya.* Nairobi: Government Printer.

———. 2003. *2003/04–2007/8 Strategic Plan for National Statistical System.* Nairobi: Central Bureau of Statistics.

Santiago, Paulo, Karine Basri E. Tremblay, and Elena Arnal. 2008. Vol. 1 of *Tertiary Education for the Knowledge Society.* Paris: Organisation for Economic Co-operation and Development.

Slaughter, Sheila, and Larry L. Leslie. 1997. *Academic Capitalism and the Entrepreneurial University.* Baltimore, MD: John Hopkins University Press.

University of Nairobi. 2005. "University of Nairobi Public Expenditure Review and Budget for the Period 2006/2007, Submitted to the Ministry of Education, Science and Technology and Commission for Higher Education." Nairobi: University of Nairobi.

Varghese, N. V. 2001. *The Limits of Diversification of Sources of Funding in Higher Education.* Paris: UNESCO International Institute for Educational Planning.

Wangenge-Ouma, Gerald. 2006. "Income Generation and the Quality Conundrum in Kenya's Public Universities." Research Report, Organization for Social Science Research in Eastern and Southern Africa, Addis Ababa.

Wangenge-Ouma, Gerald. 2008a. "Higher Education Marketization and its Discontents: The Case of Quality in Kenya." *Higher Education* 56 (4): 457–471.

———. 2008b. "Globalization and Higher Education Funding Policy Shifts in Kenya." *Journal of Higher Education Policy and Management* 30 (3): 215–229.

Wangenge-Ouma, Gerald, and Nico Cloete. 2008. "Financing Higher Education in South Africa: Public Funding, Non-Government Revenue and Tuition Fees." *South African Journal of Higher Education* 22 (4): 906–919.

Wangenge-Ouma, Gerald and Frederick Q. Gravenir. 2006 "Admission of Students in Kenyan and South African Public Universities: A Comparison." *African Journal of Education Studies* 1 (2): 30–39.

# Chapter 9

# Global Competition, Local Implications: Higher Education Development in the United Arab Emirates

*Daniel Kirk and Diane Brook Napier*

## Introduction

The United Arab Emirates (UAE) is undergoing rapid modernization in all sectors. This young nation, formed in 1971, united seven sheikdoms (Abu Dhabi, Dubai, Sharjah, Umm al-Quwain, Ajman, Fujairah, and Ras al-Khaimah) and transcended ancient rivalries and disputes to amalgamate the interests of survival and development in a volatile region. Being fortunate to have vast wealth from oil reserves, the UAE subsequently embarked on policies for accelerated modernization in all sectors, and expansion of education at all levels—including higher education, in which the trends mirror many of those in the transformation of higher education worldwide. Investment in higher education is seen as crucial for a number of purposes: for national development and global competitiveness; for the citizenry to attain improved living standards and achieve democratic participation; for meeting labor force needs and internal national development goals; and for the nation to participate in the global arena of education and economic activity. In the UAE, regional influences and development strategies are also important in shaping decisions about higher education. Within the broad landscape of global higher education transformation, the case of the UAE is distinctive as it reflects both global and regional trends and influences, and embodies several internal characteristics that impinge on higher education policy and practice.

In this chapter, we contemplate the case of the UAE with regard to higher education, and provide an overview of the educational development of the UAE, with particular reference to the meteoric rise of higher education in the last decade. It is linked to the desire of the UAE to position itself within global and regional arenas as a competitive player, and to meet internal development goals for modernization—with Western capitalist features—in a form compatible with preserving traditional Islamic society. We examine the issues related to the decisions made to import foreign higher education programs and personnel to achieve rapid growth in an emerging hybrid system, in which rapid private educational provision is being encouraged and supported by government initiatives alongside more modest expansion of federal institutions. We also consider the manner in which significant refocusing and expansion of higher education in the UAE relates to national development. We describe aspects of new policies and current practice for substantial increases in enrollments and student participation, along with the issues and implications attendant with rapid and focused growth.

We conclude by posing questions inherent in the decisions made in the UAE regarding modernization, educational policy, and higher education expansion. We look at whose interests are served, who benefits, and what aspects of social justice lie within the case. We also examine the short- and long-term prospects for sustainability of the UAE growth trajectory, for society and development in general, and for higher education in particular. Finally, we ponder the UAE as a modernizing nation, considering its position—and positioning—in both regional and global arenas in the current climate of economic adversity.

# Global Trends in Higher Education Transformation, and the Case of the UAE

The global transformation of higher education has impacted countries in all stages of development, including the UAE (Baker and Wiseman 2005, 2008; Zajda 2005; Schugurensky 2007; Zajda et al. 2009). Universities and other higher education institutions (HEIs) are undergoing restructuring, refocusing, and expansion on an unprecedented scale. Markedly increased participation reflects a shift in focus from the traditional notion of higher education occurring in exclusionist elite institutions, to programs forming part of an overall mass education system—a trend of "massification" linked to "democratization" of education systems catering to a wider spectrum of society (Baker and Wiseman 2008). Vocational and technical training programs previously viewed as inferior to "academic" higher education are enjoying a revival in recognition of their relevance for economic development and labor force needs in countries worldwide. Increasingly, countries are embroiled in intense competition, policy formation to enhance global competitiveness, and policy or program borrowing (Phillips and Ochs 2004). Notions of "knowledge production" and the desirability of creating a competitive "knowledge society" have also become prevalent, not only in post-industrial states but in developing countries that fear being left behind if they fail to keep up with global developments (Stromquist 2002). For impoverished countries in Africa and elsewhere, the perceived need to promote skills development and technical knowhow to be competitive in the global arena is an even greater challenge, given meager resources and other human resource development demands (King and McGrath 2002; Brook Napier 2005). As an emerging economy, the UAE has ample funds to pursue development and competitiveness, which generates interest—both locally and regionally—among students, educators, policy-makers, and employer organizations. Are resources devoted to higher educational expansion serving the interests of a small, elite local citizenry, or of the expatriate majority in the UAE?

To examine many of the educational issues and development within the UAE, we need to place the country within a wider regional and global context. Increasingly, tensions between privatization and state involvement in higher education and training emerge worldwide, as economic motives and market forces influence how institutions are funded, how they function, how they deal with accountability, and what graduates they aspire to produce (Schugurensky 2007). Some scholars argue that higher education is in a state of crisis, as a result of this cooperation within political and economic forces (Giroux 2007). Whether or not there is an actual crisis, the sweeping changes in higher education have implications for views of the value of higher education and the investments of human capital. For example, the unabashed expenditures in building a glittering regional center of corporate finance in Dubai, and the push for extraordinary rates of modernization in the UAE in general, make this country a key case in point regarding the role of higher education in the national development agenda of a given country.

In the dialectic of the global and the local (Arnove and Torres 2007), these global trends have impacts within countries on the nature of students, faculty, and curriculum; on broader society in terms of socioeconomic status, language, and participation of women; on regional and national development realities at the local level; and in specific programs such as teacher education in the largely centralized system of the UAE (Findlow 2005). One can also consider the case of the UAE in terms of externalization theory, which seeks to explain processes of borrowing, and the tensions between global forces and local traditions, needs, and issues within a given country (Phillips and Ochs 2004; Steiner-Khamsi 2004). In its decision to opt for a hybrid model of imported Western and traditional Islamic higher education as a shortcut to achieve widespread educational provision, the UAE is a fascinating case, particularly regarding the prospects for eventual indigenization of programs or for continued importation. In other emerging economies undergoing educational transformation (South Africa, for example), Diane Brook Napier (2005) and Jonathan Jansen (2004) have pointed out the implementation problems and mixed results of wholesale importation of programs from abroad without due regard to the receiving contexts. This public policy strategy of importing rather than building local capacity raises questions about the long-term outcomes of the importation—a North-South transfer—of higher education programs by the UAE.

Finally, given the nature of traditional Islamic society in the UAE, and the specific considerations of education for girls and women, gender issues in the UAE's higher education system are of particular importance, especially regarding the prospects for participation by women in a modernized society (Kirk and Napier 2009b). Female participation rates in the Gulf Cooperation Council (GCC) labor markets and in education are still lower than the global average (Talhami 2004; Findlow 2009). This situation is supported, in part, through conservative policies to "protect" women against entry into the workplace, functioning through generous social service provision and traditional images of women remaining in the home as mothers and wives.

# The Regional Context and the Case of the UAE

In the UAE, regional influences on trends in higher education transformation exist alongside the need for decisions to be made regarding the optimum path for modernization and development of education. Among the universal dilemmas of educational reform, Daniel Kirk and Diane Brook Napier (2009a) noted some of particular pertinence to the UAE within its unique regional context: expansion (quantity) versus quality; regional/national goals versus local internal needs; resolving questions of equity and equality (particularly regarding gender differences); developing an indigenous, homegrown system versus importing systems, programs, and qualified personnel; reconciling neocolonialism and independence; and managing continuity versus change (such as in preserving elements of traditional society while embracing modernization). These dilemmas are evident in the reform and development agenda of the UAE, which has chosen to construct a hybrid higher education system with both traditional elements and modern, "Western" imported features. Since the UAE is a member state of the Gulf Cooperation Council (along with Bahrain, Kuwait, Oman, Qatar, and Saudi Arabia), its story of higher education has to be considered in Gulf regional context, too, as follows.

An important component of the regional context in the UAE is its membership in the GCC, whose aims include improving educational provision and access at all levels in the member states through regional collaboration; promoting Gulf Arab educational opportunities; and generally promoting national and regional development, modernization, and global competitiveness. The GCC, formed in 1981, aims to foster closer collaboration and ties between its member states, and the historically linked tribal and family groups artificially separated by the creation of colonial states. The member countries have a shared cultural, religious, and historical legacy of Arab identity and culture, Islam, and a colonial legacy, as well as common economic and political interests. Among them, Saudi Arabia, the UAE, and Qatar are oil-resource rich, which make them somewhat different from most developing countries in that they have significant revenues for capacity building and development. These revenues are a key feature of the story of higher education in the UAE.

The impetus for creation of an indigenous UAE higher education system emerged as part of the overall thrust for continued modernization and independence of education in the UAE. Muhammad Morsy Abdullah (2007) notes that because schools traditionally relied on teachers from other countries in the region, these teachers became agents of change; in addition, because the ruling families sent their members abroad for university studies, the returning graduates brought further influences from abroad. These factors served as an impetus for creating a university system in the UAE, as one component of the nation's modernization and development.

The development of higher education in this region occurred in response to national development goals as well as to the rapid expansion of primary and secondary education provision across the Arabian Peninsula, influenced by the established school systems in Egypt and, later, Kuwait. University education was linked to human resource development needs. Regionally, Egypt, Jordan, and Kuwait led the way in higher education provision, having infrastructures that were comparatively well developed. In non-oil rich Egypt and Jordan, the focus was heavily on capacity building and human resource development, with higher education programs including cultural, moral, and religious dimensions. In contrast, as Daniel Kirk (2009) points out, oil-rich Saudi Arabia, Qatar, and the UAE sought to develop higher education to reduce their reliance on expatriate personnel, and to foster cultural transmission and modernization.

Educational development in the region is greatly influenced by the "ideological battle... of balancing traditional values and the contemporary scene" (Griffin 2006, 22). In the region, many are suspicious of Western imports and cultural influences, including educational systems, which are regarded as placing the traditions of Islamic cultures at risk. Acute shortages of indigenous educated and skilled workers plague each of these countries, despite significant differences in populations and demographics among the countries.

In the UAE, because of the shortage of skilled, qualified nationals and ample revenues to pay for importing workers, there is heavy dependence on expatriate or foreign workers in all sectors, including education. For example, outside of Saudi Arabia, most male teachers in Gulf national schools are Arabic-speaking expatriates on short-term contracts from Jordan, Yemen, and Egypt (Shaw 2006). Development of higher education in the Gulf Region, and UAE in particular, was gradual to begin with, with a rapid acceleration in recent decades. Recruiting, training, and keeping education professionals—both for schools and for universities—poses significant challenges for the UAE, given that the need to develop an indigenous, qualified teaching force is an important goal in the UAE and several other regional states. The UAE recognizes that teachers drawn from the national citizenry will promote the cultural, social, and historical identity of the country in ways that are effective and authentic, a task impossible for non-nationals to achieve (Kirk 2009).

# Development of the UAE Higher Education System

Having placed the United Arab Emirates within a regional and global educational context, we now turn to an exploration of the young country's lofty educational aspirations. The development of higher education in the UAE began in 1978, when a fledgling UAE government approved the formation of the United Arab Emirates University (UAEU), with a mission to "meet the educational and cultural needs of the UAE society by providing programs and service of the highest quality" (UAEU 2006, 37). The UAEU is located in Al Ain, the birth place of the founding ruler of the UAE, and an historically symbolic location in the country; it set out to replicate older, more established Arab educational institutions in Egypt and Kuwait, while simultaneously addressing the local needs of the national population. The government saw higher education as a fundamental element of developing a nation-state identity, grounded in regional and cultural experience (Kirk 2007, 2008). UAEU follows cultural norms of a segregated campus: male and female students study and live separately, often in identical buildings, while sharing academic and non-academic personnel.

As demand for university education grew, the federal government explored the establishment of other HEIs, driven in part by the interest of several rulers of Emirates in having HEIs under their own control, and in part by traditional tribal rivalries and competition among the Emirates. Thus emerged a rapidly changing and expanding higher education sector in the UAE, with the following types of institutions: (a) three federal institutions (UAE's University, Zayed University, and the Higher Colleges of Technology); (b) Emirate-funded, semi-government institutions (for example, the American University of Sharjah and Ajman University of Science and Technology); and (c) private, foreign institutions with campus operations under federal—or Emirate government—patronage and licensure (for instance, the British University in Dubai, George Mason University in Ras al Khaimah, Michigan State University in Dubai, and the Sorbonne University in Abu Dhabi).

According to the Commission for Academic Accreditation (CAA), a federal institution responsible for licensing and accrediting both national and foreign higher education providers, in the UAE there are over 60 licensed colleges and universities, offering nearly 400 accredited programs of study (CAA 2009). Following a traditional element of Arab social order, these universities are mainly clustered around education zones, areas designated for the location of higher education. Following the *souk* (market) structure of Arab commerce, many of the new HEIs are located in specifically selected locations with clustering of like services, such as University City in Sharjah, Academic City and Knowledge Village, both in Dubai, and the Abu Dhabi Education Zone, near the capital city. These "academic souks" are marketed and promoted by the central and local governments to both students and overseas education institutions, which are invited in to set up business in the UAE.

Foreign private universities often amount to little more than a scaled-down overseas presence of their home campuses, spearheading the rapid growth of foreign education in the UAE. Such a free-market driven model of developing educational provision in the country is not unexpected, given that the UAE economy is built on the market forces of tourism, commerce, business, and global financial transactions. The importation of foreign and non-indigenous systems, models, and institutions is in keeping with the prevailing perception in the Emirates (especially in Dubai and Abu Dhabi) that a rapid solution to national workforce issues and educational reform policies can be "bought-in" from overseas, in a process akin to the regional historical practice of importation of labor. This situation is a result, in part, of the relatively small number of Emiratis, who currently make up a mere 19 percent of the population (CIA 2009), the remainder consisting of expatriates, primarily from Asia and Western Europe (Kirk 2008).

A new Ministry of Higher Education and Scientific Research (UAEMHESR) was created in 1992 to oversee and accredit, through the branch of the Commission for Academic Accreditation, all of the nation's institutions, both public and private. In general, the higher education sector of the UAE has experienced rapid growth. For instance, enrollments at the UAEU climbed from 502 students in 1997–1998 to 14,741 students by 2006–2007 (International Unit 2009). These large increases in student numbers at the UAEU mirror growth in enrollments at other government and private HEIs in the country, reflecting the tremendous resources allocated to developing higher education provision in the country, which currently stands at 1.3 percent of GDP (CIA 2009). Policies in the UAE have allowed for the emergence of a complex, multifaceted higher education sector, as described above, which provides for an extraordinary array of choices and opportunities for students both from the UAE and also from other states in the region.

In the area of teacher education, there is a recognized need for the training of local teachers to reduce the dependence on foreign educators. Several universities, including UAEU, established teacher education programs that have been imported virtually intact from the United States, training UAE teachers by foreign instructors for service in local schools. In these programs, there are emergent language and cultural issues, since the language of instruction is often English, and the curriculum consists primarily of Western teaching methods and content which are not pertinent to UAE schools.

During his research on teacher education in the UAE, Daniel Kirk (2008) interviewed student teachers, who highlighted many of the issues that arise within the non-indigenous aspects of teacher education in the country. For example, Salwe, a female student in her third year of a four year teacher preparation program, spoke about many of the issues that concern young Emiratis when discussing education and the training of teachers in the UAE; her perceptions are characteristic of the sentiments expressed by other student teacher informants at several institutions:

> We need to make being a teacher good, a job Emiratis want to do and respect…. To be a teacher in UAE is very important but the universities

don't get good students or many students.... Too many foreign teachers in the [government] schools means that UAE people do not have teachers that are like them, from the same town or place and who understand what it is to be an Emirati.... The [central government] must make teaching better, with more pay and more benefits so that my friends will want to be teachers and help build our country. (Interview with Salwe 2007)

The perception of teaching as a career among the citizenry of the UAE is a mixed one. Despite respecting and promoting the role of a teacher, the government actually assigns the role of teaching to an almost exclusively imported labor force. This situation has an impact on the cultural relevance of the teaching workforce in the UAE, as Emirati teachers are a minority in their own schools, often working under the control and supervision of non-indigenous managers. This creates a dearth of role models for possible future teachers, as youngsters predominantly identify teachers as foreign workers, not as individuals who share their cultural and social heritage and understandings. Students such as Salwe feel that this situation is untenable. The implication is that the engagement of national citizens needs to be active, and rising investment in the education of young Emiratis should increase if the country is to continue to grow and become a major economic force globally—a stated policy goal of the government (UAEMOE 2008).

## Issues and Implications

Given the huge investment in developing a multifaceted, diverse higher education sector with local, regional, and international institutions— drawing students both locally and from a wide array of countries in the region—the UAE seems to be developing a well-consolidated position as a higher educational center in the region, just as Dubai has sought to position itself as a corporate and financial center in the wider region. It is too early to tell, however, whether the decisions made by the UAE will indeed produce a sound and sustainable higher education system overall. Gross enrollment in higher education in the UAE has risen dramatically over a very short period of time, from 6.7 percent in 1990 to nearly 23 percent in 2002, with projections of reaching nearly 40 percent by 2012 (World Bank 2008). Overall, the relentless pursuit of modernization and global competition by the country's many stakeholders—including students, educators, government, and institutional administrators—might well have unforeseen consequences for the UAE in the long run. Whose interests are really being served by the country's higher education programs and plans?

Do they reflect, as Henry Giroux (2008) suggests, an emergence of an "instrumentalized university," a merging of corporate culture and education in the UAE that is not in the best interests of the country's citizens? If UAE universities (especially private institutions) favor programs based on global market competitiveness, and are driven by market forces, they risk vulnerability to the instability currently rocking global financial markets, and they disavow the social justice inherent in delivering an equitable and holistic higher educational experience within a wider liberal education framework.

To this end, the UAE government has continued to state that education is a public policy priority, one that will not be adversely affected by the current economic situation. Policies, reforms and plans have been made public, partly to set out the initiatives for the public, and partly to ease possible concerns faced by overseas universities with campuses in the country, such as sustainability, funding and regional stability (Kirk 2009). These new plans include increased government support in setting up institutions, continued financial support for national students in the form of scholarships, rigorous accreditation and licensing procedures, academic integrity policies to give value to awarded degrees, and an opening up of the system to market forces (*The National* 2009).

One wonders, however, if the aggressive drive for higher education modernization in the UAE will compromise or fulfill its aspirations in gaining prominence on the regional and global stages. Still, being extraordinarily wealthy by world standards, the Gulf States, including the UAE, have vast resources to finance development (*The Economist* 2008), and their primary concern remains how to spend their wealth. Long term studies may provide answers to whether or not the UAE will attain its stated goals with regards to educational development in general, to higher education and teacher education in particular, and to education as a part of broader national and regional development.

# References

Abdullah, Muhammad Morsy. 2007. *The United Arab Emirates: A Modern History.* Abu Dhabi:
Makarem G. Trading and Real Estate LLC.
Arnove, Robert F, and Carlos A. Torres, eds. 2007. *Comparative Education: The Dialectic of the Global and Local.* Lanham, MD: Rowman and Littlefield.
Baker, David P, and Alexander W. Wiseman. 2005. *Global Trends in Educational Policy.* Amsterdam: Elsevier.
———, eds. 2008. *The Worldwide Transformation of Higher Education.* Amsterdam: Elsevier.

Brook Napier, Diane. 2005. "Implementing Educational Transformation Policies: Investigating Issues of Ideal Versus Real in Developing Countries." In *Global Trends in Educational Policy*, ed. D. P. Baker and A. W. Wiseman. Amsterdam: Elsevier.

Central Intelligence Agency. 2009. *The World Factbook*. Washington, DC: CIA. https://www.cia.gov.

Commission for Academic Accreditation. 2009. *Active Institutions*. Abu Dhabi: Commission for Academic Accreditation. http://www.caa.ae.

*The Economist*. 2008. "The Rise of the Gulf." *The Economist*, April 24, 15. http://www.economist.com.

Findlow, Sally. 2005. "International Networking in the United Arab Emirates Higher Education System." *Compare* 35 (3): 285–302.

———. 2009. "The Role of Education in Meeting Development Requirements." Paper Presented at the 14th ECSSR Annual Conference, Abu Dhabi, February 22–26, 2009.

Giroux, Henry A. 2007. *The University in Chains: Confronting the Military-Industrial-Academic Complex (The Radical Imagination)*. Boulder, CO: Paradigm Publishers.

Griffin, Rosarii, ed. 2006. *Education in the Muslim World: Different Perspectives*. Oxford: Symposium Books.

International Unit. 2009. *United Arab Emirates*. London: International Unit. http://international.ac.uk.

Jansen, Jonathan D. 2004. "Importing Outcomes-Based Education into South Africa: Policy Borrowing in a Post-Communist World." In *Educational Policy Borrowing: Historical Perspectives*, ed. D. Phillips and K. Ochs. Oxford: Symposium Books.

King, Kenneth, and Simon McGrath. 2002. *Globalisation, Enterprise and Knowledge: Education, Training and Development in Africa*. Oxford: Symposium Books.

Kirk, Daniel. 2007. "Local Voices, Global Issues: Teacher Education in the UK, USA and UAE." Paper presented at the CIES Annual Conference, Baltimore, MD, March 2007.

———. 2008. "Local Voices, Global Issues: A Comparative Study of the Perceptions Student Teachers Hold in Relation to Their Pre-service Education in the United States of America, England, and United Arab Emirates." PhD Diss., College of Education, University of Georgia, Athens.

———. 2009. "If We Build It, They Will Come: Educational Borrowing and Higher Education Development in the United Arab Emirates." Paper presented at the Comparative and International Education Society Annual Conference, Charleston, South Carolina, 22 March 2009.

Kirk, Daniel, and Diane Brook Napier. 2009a. "The Transformation of Higher Education in the United Arab Emirates: Issues, Implications, and Intercultural Dimensions." In *Nation-Building, Identity and Citizenship Education: Cross Cultural Perspectives*, ed. J. Zajda, H. Daun, and L. J. Saha. Dordrecht, The Netherlands: Springer.

———. 2009b. "Issues of Gender, Equality, Education, and National Development in the United Arab Emirates." In Vol. 10 of *Gender, Equality, and*

*Education from International and Comparative Perspectives*, ed. D. P. Baker and A. W. Wiseman. Amsterdam: Elsevier.

*The National.* 2009. "Minister puts universities in hand of market forces." *The National*, May 20. http://www.thenational.ae.

Phillips, David, and Kimberly Ochs, eds. 2004. *Educational Policy Borrowing: Historical Perspectives.* Oxford: Symposium Books.

Schugurensky, Daniel. 2007. "Higher Education Restructuring in the Era of Globalization: Toward a Heteronomonous Model?" In *Comparative Education: The Dialectic of the Global and the Local*, ed. R. A. Arnove and C.A. Torres. Lanham, MD: Rowman and Littlefield.

Shaw, Ken E. 2006. "Muslim Education in the Gulf States and Saudi Arabia: Selected Issues." In *Education in the Muslim World: Different Perspectives*, ed. R. Griffin. Oxford: Symposium Books.

Steiner-Khamsi, Gita, ed. 2004. *The Global Politics of Educational Borrowing and Lending.* New York: Teachers College Press.

Stromquist, Nelly P. 2002. *Education in a Globalized World: The Connectivity of Economic Power, Technology, and Knowledge.* Lanham, MD: Rowman and Littlefield.

Talhami, Ghada Hashem. 2004. *Women, Education and Development in the Arab Gulf Countries.* Abu Dhabi: Emirates Center for Strategic Studies and Research.

United Arab Emirates' Ministry of Education. 2008. "MOE Strategic Objectives." Abu Dhabi: Ministry of Education. http://www.moe.gov.ae.

United Arab Emirates University. 2006. "Student Teacher Handbook." Al Ain, UAE: United Arab Emirates University.

World Bank. 2008. *The Road Not Traveled. Education Reform in the Middle East and North Africa.* Washington, DC: World Bank.

Zajda, Joseph, Holger Daun, and Lawrence J. Saha, eds. 2009. *Nation Building, Identity, and Citizenship: Cross Cultural Perspectives.* Dordrecht, The Netherlands: Springer.

Zajda, Joseph, ed. 2005. *The International Handbook of Globalisation and Educational Policy Research.* Dordrecht, The Netherlands: Springer.

# Chapter 10

## Global Competition as a Two-Edged Sword: Vietnam Higher Education Policy

*Diane E. Oliver and Kim Dung Nguyen*

## Introduction

From the perspective of a developing country, in this case Vietnam, global competition in higher education is a two-edged sword. On one side are entities from exporting countries competing to provide educational services that could increase the capacity and quality of a higher education system constrained by limited financial and human resources. On the other are factors that could negatively affect the quality of higher education in Vietnam, such as unscrupulous foreign education providers. Thus, the sword can cut in two directions, or, as stated in a Vietnamese voice, "We can enjoy fresh winds to make our health better; concurrently, there may be unhealthy winds that do harm to us with weak bodies" (Nguyen et al. 2006, 2).

Two dynamics have been heightening the visibility of global competition in higher education as it plays out in Vietnam: (a) accession to membership in the World Trade Organization (WTO), and (b) increases in cross-border education activity. In this chapter, the necessity for Vietnam to establish policies that address global competition in higher education is analyzed by examining the implications of WTO and cross-border education. However, before moving into the analysis, both a conceptual framework and Vietnam's higher education context are discussed.

# Globalization as a Framework for Understanding Global Competition

Most scholars who currently study globalization seem to agree that it is a set of processes that make borderless the important economic, social, and cultural practices previously bounded within nation-states (Suarez-Orozco and Qin-Hilliard 2004). Globalization is characterized by the flow of economic, technical, and intellectual trends across permeable borders, and the tendency of each country to respond differently as a result of their unique histories, cultures, and political structures (Knight 1999; UNESCO 2004).

Although a world economy has existed since at least the sixteenth century, when inter-oceanic trading systems were established (Coatsworth 2004), the current global economy has been made possible by an advancing technological infrastructure, including telecommunications, information systems, and transportation (Carnoy 2005). Concomitantly, knowledge has become both more portable and increasingly important to global markets and the wealth of nations. Thus, globalization creates increased pressure on countries to provide more higher education, and higher quality within that system. Moreover, according to David Bloom (2002), "globalization has turned a piercing spotlight on each country's higher education system and institutions" (2). Consequently, the forces of globalization have given rise to the worldwide phenomenon of increased global competition in higher education.

# Vietnam's Higher Education Context

Vietnam has a population of over 86 million, and according to Long S. Le (2007), approximately 26 million are between the ages of 15 and 30. The official literacy rate as of 2006 was 90 percent (CIA 2008). In 2007, the economy grew by 6.23 percent (U.S. Department of State 2009), and the per capita income was US$1,204 (U.S. Department of State 2009). Today's Vietnam, in many ways, is the result of a national policy adopted in 1986 by the Sixth Party Congress called *Doi Moi* (Renovation). This policy required major changes to move from a centralized five-year plan to a socialist market economy. In the higher education system, two important changes were the implementation of tuition, and diversification in types of institutions; five types of higher education institutions (HEIs) are currently in existence: public, people-established, semipublic, foreign-owned, and private.

There are currently 332 universities and colleges in Vietnam (which is double the number existing in 1999), and the number of faculty has risen since then by approximately 43 percent (VMOET 2008a). As of 2007, there were over 1.5 million students in the higher education system. Yet, according to Michael Marine (2007), a former U.S. Ambassador to Vietnam, 1.8 million prospective students sat for the university entrance examination in July 2007, competing for just 300,000 new openings. Although the number of students admitted to public universities after taking the national entrance examination has increased from 10 percent to 17 percent (VMOET 2008b), access continues to be a challenge.

The Ministry of Education and Training (VMOET) has primary authority for governance of the higher education system in Vietnam. Curriculum is currently based on a structure of 60 to 70 percent (depending on disciplines) stipulated in detail by VMOET (the core), and 30 to 40 percent stipulated as an outline by VMOET that allows for the addition of some subjects based on local needs. However, the Resolution No. 14/NQ-CP (November 2, 2005) on *Fundamental and Comprehensive Higher Education Reform in Vietnam 2006–2020* proposes to provide increased legal autonomy for institutions, while simultaneously developing a quality assurance and higher education accreditation system.

# GATS and Higher Education: Implications for Global Higher Education

On January 11, 2007, Vietnam became the 150th member of the WTO, which automatically stipulated compliance with certain key provisions of the General Agreement on Trade in Services (GATS). Compliance with other provisions of GATS depends on the extent of additional promises made by a country on a negotiated document referred to as the country's "Schedule of Specific Commitments" (Terry 2006). Thus, although GATS is described as being voluntary because countries determine which sectors are placed on their national schedule of commitments, some aspects of the agreement are mandatory, such as progressive liberalization of trade for items that appear on the schedule (Knight 2004). As Jane Knight (2002b) asserts, "GATS is not a neutral agreement. It aims to promote and enforce the liberalization of trade in services" (9).

Education is one of 12 service sectors within GATS, and is further divided into five service subsectors: primary, secondary, higher, adult and continuing, and other. For higher education, GATS's purpose of promoting freer trade means creating increased market access and competition in global higher education by removing barriers such as licensing requirements and lack of opportunity for accreditation (Verger 2009) and facilitating the cross-border movement of higher education services. As of 2006, 48 countries have made commitments to education in GATS, of which 39 have included commitments to trade liberalization in the higher education subsector (Raychaudhuri and De 2007).

GATS specifies four modes of trading services in the 12 sectors: (a) cross-border supply where the service crosses the border (e.g., distance education); (b) consumption abroad, where the customer (e.g., the student) goes to the supplier's country; (c) commercial presence when the supplier has facilities in another country to provide services (e.g., branch campuses); and (d) presence of natural persons where an individual (i.e., a professor) travels from another country to provide service temporarily (Verger 2009). Moreover, as explained previously, the GATS framework includes general (unconditional) obligations that apply to all WTO members, as well as specific commitments (conditional obligations) that are negotiated by member countries.

One of the unconditional obligations under GATS that is of particular importance to increased global competition in higher education is that of most favored nation (MFN) status. MFN status requires equal treatment of all service suppliers from WTO member countries; thus, if Vietnam allowed or denied a specified level of foreign competition in higher education services for one WTO trading partner, the same opportunity, or exclusion, would apply to all WTO members.

Conditional obligations include national treatment and market access. National treatment requires that foreign and domestic providers receive equal treatment, while market access is the degree to which foreign providers are permitted to enter a country's domestic market in a specific service sector; the countries can limit or expand their market access, but these commitments must be listed on the "Schedule of Specific Commitments." Limitations could include restrictions on the total number of foreign suppliers and operations, number of natural persons employed, transaction values, and foreign capital participation (Varghese 2009). However, although WTO members can tailor their commitments as desired, during each new negotiation period countries are expected to include more sector commitments and reduce limitations on national treatment and market access (Verger 2009). With the rise of global competition in higher education throughout the world due to escalating demand, market forces, and growth in income generating cross-border activities, such as overseas campuses (Altbach and Peterson 2008) increased pressure to liberalize trade in this service subsector is likely to occur.

Debate over the impacts of including higher education in GATS has been intense, and nearly polarized advocates on either side. Knight (2002) suggests that the issues revolve around two key questions: "Trade liberalization 'for whose benefit' or 'at what cost'" (6). On one side of the heated discussions are those who support the inclusion of higher education in GATS by arguing for the benefits of greater access, competitiveness, diversification of providers and delivery modes, and financial gain (Bubtana 2007). Proponents in developing countries view increased global competition in higher education through trade liberalization as an opportunity to have foreign providers help address the problem of limited student access to higher education, and develop the country's educational infrastructure.

On the other side of the debate are those who argue that the inclusion of higher education in GATS would change it from a public good to a tradable commodity in the global market. N. V. Varghese (2007) asserts that trade is profit-motivated, while the essence of education is not; thus, trade policies resulting from GATS may conflict with national educational interests. Philip Altbach (2001) similarly expresses concern regarding the impact of GATS on developing countries: "Once universities in developing countries are subject to an international academic marketplace regulated by the WTO, they would be swamped by overseas institutions and programs intent on earning a profit but not contributing to national development" (para. 13). Some developing countries fear there will be high social costs associated with cross-border providers, including an elitist education system, further reduction in government support for education, and ultimate disappearance of a national higher education system due to an inability to compete with foreign education service providers (Bubtana 2007). Moreover, developing countries have quality assurance concerns, as they lack the capacity to protect themselves from diploma mills and poor quality foreign providers (Verger 2009).

# GATS and Higher Education: Implications for Vietnam

Given the core concept of GATS (progressive trade liberalization) and the debated implications of trade liberalization in global higher education competition, Vietnam must carefully consider its higher education policy needs. As stated by Frank Newman and Laura Couturier (2002), "Competition offers higher education the opportunity to fix stubborn problems. However…the dangers are great. The key is finding policy solutions that help steer the market in ways that benefit society and serve the greater public good" (para. 2).

Vietnam should begin its preparations for future GATS trade negotiations by conducting policy analysis that addresses the current opportunities and threats related to cross-border education operations, which are fueled by global competition, and reflect arguments on both sides of the debate over including higher education in GATS. Thus, the challenges and responses to global higher education competition in the form of cross-border education are analyzed in the remainder of this chapter. Our analysis and related recommendations aim to assist Vietnam's policy makers, educators, and trade negotiators, who must work together when assessing future GATS commitments in the higher education subsector.

## Cross-Border Higher Education

UNESCO (2004) characterizes cross-border education as the movement of education across jurisdictional boundaries, with the nation retaining its regulatory responsibility, particularly in the areas of quality, access, and funding. Moreover, cross-border education includes the movement of students, faculty, knowledge, educational programs, curriculum, and providers from one country to another (OECD and World Bank 2007). Cross-border education is a multibillion dollar industry (Lee 2005), and cross-border higher education is on the rise, thus creating competition among domestic and foreign institutions.

However, "at the heart of much debate" is concern about quality and accreditation (Knight 2002b, 22). All types of education products can be freely exported, and it has become difficult to regulate trade in the area of HEIs, programs, degrees, and certificates. A related concern for the public is confusion over the value of qualifications offered by imported programs and their acceptability to the labor market (UNESCO 2004). Thus, the challenge is how to maximize the benefits and minimize the risks associated with liberalizing trade in higher education. One important step is to develop structures for quality assurance and accreditation systems that address foreign providers and programs (UNESCO 2005).

## Cross-Border Higher Education in Vietnam

The analytical framework used in the following section to examine the implications of cross-border higher education in Vietnam is comprised of four categories: policy, legislative and regulatory, cross-border educational, and financial. While each is discussed separately, it should be noted that the categories are interconnected.

## Policy Implications

The forces, processes, and demands of global competition make it necessary for Vietnam to develop a strategic plan that defines the country's goals, objectives, and a timetable for strengthening its higher education system, first domestically and then to compete internationally. Raising concerns about quality, accreditation, diversification, and other related issues is important preparation for making informed policy decisions that will improve the higher education system.

The government of Vietnam has issued a reform agenda that envisions a bigger and better national higher education system by 2020. For example, VMOET announced an intention to train 20,000 PhDs by 2020 (*Wasley* 2007), with 10 percent of them to be sent abroad for doctoral degrees (*Viet Nam News* September 16, 2008). In this case, cross-border education presents an opportunity to increase the quality of faculty in Vietnam; yet a potential threat is that faculty, researchers, or graduating students will stay abroad or move to a more developed country for better salaries and working conditions. Moreover, there are demographic and higher education trends in the United States and other developed countries that suggest rising competition for college educated international workers, particularly in science and technology (Galama and Hosek 2009). However, the essential point is that effective policy requires strategic planning, which involves more than a vision; it must specifically factor in the opportunities and threats of cross-border education in developing Vietnam's higher education capacity, quality, and international recognition.

## Legislative and Regulatory Implications

An important concern is the way in which cross-border education providers are to be dealt with in Vietnam's *Education Law*. Because cross-border education is relatively new in Vietnam, it is not clearly conceptualized; and in a country where HEIs are only authorized to do what the regulations specifically address, the absence of regulations, or having regulations that are too vague, is troublesome. Currently, the country's *Education Law* has one regulation addressing two components of cross-border education, and it is quite general: (a) foreigners and overseas Vietnamese are encouraged to teach, study, collaborate, use, and transfer scientific work in Vietnam; and (b) collaboration between open educational institutions in Vietnam and overseas Vietnamese or foreigners must comply with government regulations.

We believe more specific laws, regulations, and oversight must be developed now to prevent abuses by those who, as stated by Mark Ashwill (2005), see private and cross-border education "as a lucrative market to be tapped and a valuable commodity to be sold to the highest bidder" (69). Regulatory structures that address the diversity of providers and delivery modes introduced by cross-border education, as well as increases in international trade, must be developed. As stated by Knight (2002a), such a structure links regulations to national policy and protection of the public through "licensing, regulating, [and] monitoring both private (profit and non-profit) and foreign providers" (7).

Vietnam has already experienced the negative consequences of not being able to monitor the quality of cross-border higher education programs. For example, Ashwill (2005) reports that the Taiwan Asian International University, established in 1995 through cooperation with a Vietnamese university, was discovered in 2000 to be a hoax; over 2,000 students and their families lost "hundreds of thousands of dollars" (68). As this devastating example makes clear, it is imperative that higher education programs which link domestic and foreign HEIs in Vietnam be brought under government control. However, the situation is complex, as the government has in recent years expressed a need to move toward increased territorial decentralization that distributes control vertically to the provincial, local, and institutional levels (Oliver et al. 2009). The Minister of Education and Training recently stated that institutions will be given more autonomy and because VMOET is already constrained with so many responsibilities, decentralization to the provinces and localities is necessary (*Viet Nam News* August 26, 2009). Also, major international donors, (such as the World Bank and Asia Development Bank) have pressed for diversification in types of higher education institutions and funding sources (Oliver et al. 2009). Thus, the challenge is to regulate, but not over regulate.

As of now, regulations regarding registration and ongoing supervision of education programs offered by foreign entities have not been clearly specified. In addition, a specialized unit has not been established to monitor foreign providers and develop timely government management policies. Fortunately, Vietnam can look to other countries with experience in this regulatory field—including Hong Kong, which implemented the *Non-local Higher and Professional Education Ordinance in 1997* (McBurnie and Ziguras 2001) and China, which has regulations requiring that foreign providers go through a multistep process and partner with a domestic institution (Powar 2005). Vietnam could borrow and adapt regulatory regimes from this neighboring country. Moreover, given that importing countries have a right to request quality guarantee regimes from countries operating education services in their region (OECD 2003), Vietnam should require foreign providers to be accredited by a government-recognized accrediting body in their home country to protect its students.

Many countries require that the foreign provider be accredited in their own country before authorizing entry to the host country; for example, China requires that the provider be accredited in both countries. The basic premise of China's approach, which may be suitable for Vietnam, is that cross-border education should be used as a means for improving the quality of the national higher education system (Varghese 2007), by strengthening programs and instructional techniques and augmenting funding sources in domestic institutions through partnerships with foreign providers. However, Vietnam's quality accreditation system has only been in operation since 2004, and lack of experience and training with education quality assurance and assessment systems have caused many difficulties in this area, including administrators of institutions not knowing how to conduct self-studies or present appropriate evidence to support claims of having met accreditation criteria (Nguyen, Oliver, and Priddy 2009). Moreover, the lack of transparency regarding education quality and reasonable consumer protection regulations has created conditions favorable to some foreign providers who supply low quality programs.

Judith S. Eaton, President of the Council for Higher Education Accreditation in the United States, and Stamenka Uvalic-Trumbic of the Division of Higher Education in UNESCO (2008) note that degree mills, which are fraudulent higher education providers, have become an international concern. In Vietnam, the absence of reliable information systems makes it difficult for the public to assess and confirm the credibility and quality of cross-border education providers. Hong Kong's *Non-local Higher and Professional Education Ordinance*—mentioned above—could serve as a model for Vietnam in addressing these issues. This ordinance stipulates that cross-border education providers must apply for registration, and, in the process, provide detailed information about their courses, delivery methods, student requirements, staff, and facilities. In addition, information about foreign provider courses and the local partners is posted on the Non-local Courses Registry website (McBurnie and Ziguras 2001).

In summary, Vietnam needs policies and clear regulations that will strengthen its quality assurance structure by: (a) addressing cross-border education, (b) protecting students from poor quality education programs, (c) establishing a process for registering foreign providers, and (d) establishing a centralized data base providing information on higher education programs recognized by the government.

## Cross-Border Educational Implications

One of the concerns related to cross-border education is that providers will only offer courses in a limited number of subjects that have potential for profit such as English, business, and information technology. The implications for Vietnam are that these providers will compete with underfunded public universities that are left to offer unprofitable courses essential to the country's social and cultural development including the arts, humanities, and social sciences (Nguyen 2007). Thus, as stated by A. Bubtana (2007), the foreign curriculum may not be relevant to the sociocultural context of the importing country, which is at odds with the national mission of higher education "to preserve and promote national cultures, instil [*sic*] cultural identity and educate for citizenship" (para. 4). Another concern, which was voiced by K. B. Powar (2005) in his analysis of potential GATS impacts on India, is that cross-border education—especially when given impetus by GATS—could impact employment, as affluent students can benefit from higher quality foreign education while the majority of students from middle and lower economic class families could only access "mediocre" education funded by the government, thus resulting in a new (postcolonial) form of elitism.

Yet, cross-border education also presents important opportunities for further development of higher education. Because foreigners do not commonly speak Vietnamese, it is important for Vietnamese people to learn other languages. Thus a variety of foreign languages should be used for instruction, including English and Chinese, to support the growth of business opportunities in Vietnam. As stated by Varghese (2007), "one of the attractions of cross-border education is the socialization with an international language which has value on the labour market" (18). Also, cultural exchanges can be enhanced through cross-border education in order to prepare for diversification, which is an essential component of the globalization and integration process.

## Financial Implications

The financial implications of cross-border education are important to take into consideration as well, given that foreign providers have contributed funding that advances the development of Vietnam's higher education infrastructure. For example, Japan and the Asia Development Bank plan to provide US$1 million in 2009 to conduct a feasibility study for establishing model science and technology research universities in Hanoi and Da Nang. This study intends to assist with designing a framework of policy and regulation to further the Government of Vietnam's "aim to make higher education more innovative, internationally competitive, and relevant to labour market needs" (*Viet Nam News* September 1, 2008, 2). In addition, a German-Vietnam university was opened in September 2008 with the assistance of EUR 450,000, with the intention of helping to "promote the German language and culture in Viet Nam, contributing to the exchange of students and lecturers, as well as bringing long-term benefits to both sides" (*Viet Nam News* September 11, 2008, 4).

While these types of cross-border education projects represent opportunities in helping to build the higher education infrastructure, they also raise concerns about negative impacts on the public universities. For example, one of Vietnam's most challenging problems is a shortage of faculty, and it is likely that cross-border higher education providers would pay larger salaries, thereby drawing highly qualified faculty away from public institutions; alternatively, the higher salaries might cause the public institution's faculty members to moonlight, thus creating work overloads that adversely affect the quality of teaching. Moreover, there is concern that when the government commits funds to cross-border education projects, funding for the country's public HEIs may decline or remain flat (Nguyen 2009).

# Recommendations

Based on the analysis of potential GATS implications and concerns regarding cross-border education in Vietnam today, some recommendations can be made. These recommendations indicate that responsibilities and actions taken at the government and HEI levels must be connected and coordinated.

First, the government should establish an effective, transparent, comprehensive, and fair procedure for licensing foreign HEIs to operate in Vietnam, including online training programs. The processes established by China and Hong Kong could be a starting point for developing Vietnam's procedures. Concurrently, Vietnam needs to develop and implement regulations, guidelines, and appropriate measures to protect students from participating in low quality programs provided by domestic and foreign providers. Adaptation of Hong Kong's course registry model would be useful in conceptualizing this effort. Moreover, Vietnam needs to leverage the benefits of cross-border education through establishing appropriate policies to minimize the potential "brain drain" associated with study abroad. Research should be conducted to identify the extent to which brain drain is occurring and to analyze the specific causes; however, this type of evaluation research is not useful unless the cycle of action and assessment is implemented.

Second, a cadre of quality assurance experts should be trained to expedite the implementation of quality accreditation in Vietnam, and an independent education accreditation council should be established to ensure the conformity of accreditation standards and regulations throughout the higher education system. Moreover, accreditation should be expanded to include private as well as public institutions, and the addition of accreditation standards or criteria that address the quality of cross-border partnership programs should be examined. International cooperation in ensuring and accrediting higher education quality should be pursued; these links will facilitate qualification recognition and credit transfer. In addition, this cooperation would enable the government to improve public access to current and accurate data on recognized foreign education providers and request the assistance of appropriate foreign agencies in determining the quality of new cross-border education entities (Knight 2002b; UNESCO 2005).

Third, there is a need to enhance transparency in cooperation and competitiveness among Vietnamese and foreign institutions. UNESCO (2005) guidelines specifically suggest transparency in both financial status and the academic programs that are offered. Information on HEIs in Vietnam must be improved and updated in an accurate, comprehensive way. Cross-border providers also must assume responsibility for ensuring the quality of information provided concerning their own and multination education programs. Increased transparency also will assist employers in understanding the qualifications of graduates.

# Conclusion

Cross-border education offers both opportunities and challenges for Vietnam. By ensuring high quality, equitable access, and the availability of programs that substantially benefit both students and the country itself, the opportunities can be realized in improving higher education. However, if cross-border education is not carefully monitored for quality assurance and relevance, it represents a threat to the country's efforts toward improving higher education and gaining international recognition. This double-edged sword of global higher education competition presents major challenges in achieving a balance between government control and progressive autonomy. Vietnam needs to identify governmental processes that will protect its citizens from dishonest providers while simultaneously decentralizing decision making in higher education management and encouraging quality cross-border higher education providers to operate in Vietnam. A critical linchpin for achieving this balance is to strengthen higher education quality assurance policies and processes along with the accreditation system. Yet, the Minister of Education and Training has stated that VMOET is already overburdened with responsibilities; therefore, new models should be explored, such as an independent accreditation body with a broadened scope for quality improvement.

While the analysis of benefits and risks associated with cross-border higher education are helpful in preparing for GATS negotiations, Verger (2009) raises another important concern. He argues that GATS negotiations are based on national interests, rather than free trade motivations. Moreover, the trade representatives of member countries make the decisions and agreements at the GATS subsystem levels. These representatives normally report to the Ministry of Trade, which often does not communicate with other government stakeholders regarding GATS decisions. He further argues that higher education may be a pawn in negotiations for other trade commodities that are considered more important by economists and trade specialists. Verger's perspectives should be cautionary to Vietnam in ensuring that VMOET and other higher education stakeholders are given a voice that is factored into negotiations affecting the higher education subsector.

In summary, WTO membership is expediting the opening of Vietnam to world trade, GATS negotiations are on the horizon, and cross-border higher education has already arrived. Global competition in Vietnam's higher education system must be addressed with well analyzed, planned, and coordinated micro-context responses to ensure that its reforms best serve the country's higher education and national interests.

# References

Altbach, Philip G. 2001. "Higher Education and the WTO: Globalization Run Amok." *International Higher Education* 23 (Spring): 2–4.

Altbach, Philip G., and Patti McGill Peterson. 2008. "America in the World: Higher Education and the Global Marketplace." In *The Worldwide Transformation of Higher Education, International Perspectives on Education and Society, Volume 9*, ed. D. P. Baker and A. W. Wiseman. Bingley, UK: Emerald.

Ashwill, Mark A., with Diep Ngoc Thai. 2005. *Vietnam Today: A Guide to a Nation at a Crossroad.* Boston: Intercultural Press.

Bloom, David E. 2002. "Mastering Globalization: From Ideas to Action on Higher Education Reform." Paper presented at a conference entitled "Globalisation: What

Issues Are at Stake for Universities?" University of Laval Conference, Quebec, September 19, 2002.

Bubtana, A. R. 2007. "WTO/GATS: Possible Implications for Higher Education and Research in the Arab States." Paper presented at the United Nations Educational, Scientific and Cultural Organisation Regional Seminar, Rabat, Morocco, May 24, 2007.

Carnoy, Martin. 2005. "Globalization, Educational Trends and the Open Society." Paper presented at the Open Society Institute Education Conference, Budapest, January 2, 2005.

Central Intelligence Agency. 2008. *The World Factbook*. Washington, DC: Central Intelligence Agency. https://www.cia.gov.

Coatsworth, John H. 2004. "Globalization, Growth, and Welfare in History." In *Globalization Culture and Education in the New Millennium*, ed. M. M. Suarez-Orozco and D. B. Qin-Hilliard. Berkeley: University of California Press.

Eaton, Judith S. and Stamenka Uvalic-Trumbic. (2008). "Degree Mills: The Impact on Students and Society." *International Higher Education* 53 (Fall): 3–5.

Gama, Titus and James Hosek. 2009. "Global Competitiveness in Science and Technology and the Role of Mobility." In *Higher Education on the Move: New Developments in Global Mobility*, ed. E. Bhandari and S. Laughlin. New York: Institute for International Education.

Knight, Jane. 1999. "Internationalisation of Higher Education." In *Quality and Internationalization in Higher Education* (pp. 13–28), ed. Organisation for Economic Co-operation and Development. Paris: OECD.

_____. 2002a. "The Impact of Trade Liberalization on Higher Education: Policy Implications." Paper presented at the University of Laval Conference on Globalisation, Quebec, September 20, 2002.

_____. 2002b. *Trade in Higher Education Services: The Implications of GATS*. Paris: UNESCO. http://www.unesco.org.

_____. 2004. "Cross-border Education in a Trade Environment: Complexities and Policy Implications." In *Proceedings of Accra Workshop on GATS: The Implications of WTO/GATS for Higher Education in Africa*, Accra, Ghana, April 27–29, 2004.

Le, Long S. 2007. "Vietnam's Generation Split." *Asia Times*, June 23. Hong Kong: Asia Times Online, Ltd. http://www.atimes.com.

Lee, Mollie N. 2005. "Cross-border Education in Asia and the Pacific Region: International Framework for Qualifications." In *UNESCO Forum Occasional Paper Series no. 8*: 100–109. Paris: United Nations Educational, Scientific and Cultural Organisation.

Marine, Michael W. 2007. "Challenges of Higher Education in Vietnam: Possible Roles for the United States." Hanoi: Embassy of the United States, Vietnam. http://vietnam.usembassy.gov.

McBurnie, Grant, and Christopher Ziguras. 2001. "The Regulation of Transnational Higher Education in Southeast Asia: Case Studies of Hong Kong, Malaysia and Australia." *Higher Education* 42 (1): 85–105.

Newman, Frank, and Laura K. Coutureir. 2002. *Trading Public Good in the Higher Education Market*. Zagreb: The Observatory on Borderless Higher Education. http://www.pravo.hr.

Nguyen, Hien T. (2007). "The Impact of Globalisation on Higher Education in China and Vietnam: Policies and Practices." Paper presented at the University of Salford Conference on Education in a Changing Environment, Manchester, UK, September 12–14, 2009.

Nguyen, Kim Dung, Diane Oliver, and Thanh Xuan Pham. 2006. "Effectiveness of Quality and Commerce: The Interaction between Integration and Quality Standards of Higher Education in Vietnam." Paper presented at the Ministry of Education and Training Sponsored Forum on WTO Entry and Higher Education Reform, Hanoi, December 12, 2006.

Nguyen, Kim Dung. 2009. "Survey on the Actual Situation of Vietnam Educational Development in the Context of Free Market and in Front of the Requirements of Globalization." Paper presented at the Government Office of the Vietnam Educational System in the Context of Free Market and Globalization conference, Hanoi, July 14, 2009.

Nguyen, Kim D., Diane E. Oliver, and Lynn E. Priddy. 2009. "Criteria for Accreditation in Vietnam's Higher Education: Focus on Input or Outcome?" *Quality in Higher Education* 15 (2): 123–134.

Oliver, Diane E., Xuan Thanh Pham, Paul A. Elsner, Thi Thanh Phuong Nguyen, and Quoc Trung Do. 2009. "Globalization of Higher Education and Community Colleges in Vietnam." In *Community College Models: Globalization and Higher Education Reform*, ed., R. L. Raby and E. J. Valeau. New York: Springer.

Organisation for Economic Co-operation and Development. 2003. *Enhancing Consumer Protection in Cross-border Higher Education: Key Issues Related to Quality Assurance, Accreditation and Recognition of Qualifications.* Paris: OECD.

Organisation for Economic Co-operation and Development and The World Bank. 2007. *Cross-border Tertiary Education: A Way Towards Capacity Development.* Paris: OECD.

Powar, K. B. 2005. "Implications of WTO/GATS on Higher Education in India." Paper presented at the UNESCO Regional Seminar on the Implications of WTO/GATS on Higher Education in Asia and the Pacific, Seoul, April 27–29, 2005.

Raychaudhuri, Ajitava, and Prabir De. 2007. "Barriers to Trade in Higher Education Services: Empirical Evidence from Asia-Pacific countries." *Asia-Pacific Trade and Investment Review* 3 (2): 67–88.

Socialist Republic of Vietnam. 2005. *Resolution No. 14/2005/NQ-CP On Fundamental and Comprehensive Higher Education Reform in Vietnam for the Period of 2006–2020,* November 2, 2005.

Suarez-Orozco, Marcelo M., and Desiree B. Qin-Hilliard. 2004. "Globalization Culture and Education in the New Millennium." In *Globalization Culture and Education in the New Millennium*, ed. M. M. Suarez-Orozco and D. B. Qin-Hilliard. Berkeley: University of California Press.

Terry, Laurel S. 2006. "Current Developments Regarding the GATS and Legal Services: The Hong Kong Ministerial Conference and the Australian Disciplines Paper." *The Bar Examiner* 75 (1): 26–34.

UNESCO. 2004. *Higher Education in a Globalized Society: UNESCO Education Position Paper.* Paris: UNESCO. http://unesdoc.unesco.org.

_____. 2005. *Guidelines for Quality Provision in Cross-border Higher Education.* Paris: UNESCO. http://unesdoc.unesco.org.

U.S. Department of State. 2009. *Background Note: Vietnam.* Bureau of East Asian and Pacific Affairs. Washington, DC: U.S. Department of State. http://www.state.gov.

Varghese, N. V. 2007. *GATS and Higher Education: The Need for Regulatory Policies.* Paris: UNESCO.

Varghese, N. V. 2009. "GATS and Transnational Mobility in Higher Education." In *Higher Education on the Move: New Developments in Global Mobility,* ed. E. Bhandari and S. Laughlin. New York: Institute for International Education.

Verger, Antoni. 2009. *WTO/GATS and the Global Politics of Higher Education.* New York: Routledge .

*Viet Nam News.* 2008. "Japan, ADB Help Develop Education." *Viet Nam News,* September 1: 2.

_____. 2008. "New University to Boost German Cooperation." *Viet Nam News,* September 11: 1 and 4.

_____. 2008. "Education Sector Gets $24 Billion." *Viet Nam News,* September 16: 1–2.

_____. 2009. "Universities Urged to Meet International Standards." *Viet Nam News,* August 26: 1 and 4.

Vietnam's Ministry of Education and Training (VMOET). 2008a. *So lieu thong ke giao duc* [Statistics on Education]. Hanoi: VMOET. http://edu.net.vn/thongke/dhcd.htm#truong.

_____. 2008b. "Summary Report of MOET for the 2006–2007 Academic Year." Report presented at a conference on the National University Entrance Examination, Ho Chi Minh City, January 5, 2008.

Wasley, Paula. 2007. "Vietnamese Leaders Discuss Overhaul of Higher Education during US Visit." *Chronicle of Higher Education* 53 (43): A41.

# Chapter 11

# Vietnam, Malaysia, and the Global Knowledge System

*Anthony Welch*

## Introduction

Scholars of higher education in North America and Europe pay little attention to Southeast Asia, even though it serves as an important source of international students in North America, the United Kingdom, and Australia. In terms of international student flows, this relative disregard is a sin of omission: Malaysia, for example, is not only among the top ten source countries for the United States, the United Kingdom, and Australia, but its own ongoing effort to position itself as an "Eduhub" has boosted the international competitiveness of its higher education system, which now attracts a large and growing number of international students.

The lack of interest in Southeast Asia by Western scholars reflects the dominance of the global knowledge system by the global North, whose scholars' research gaze rarely extends beyond their national horizons (Altbach 2003). To the extent that the global South figures at all in Western scholarship, much greater attention is paid to China and India—the two giant peripheries of the global knowledge system (Altbach 1998). In part, this relative neglect reflects the enormous diversity of Southeast Asia, which makes any generalization perilous. The ten member states of the Association of Southeast Asian Nations (ASEAN) are home to 575 million people, an average annual gross domestic product (GDP) growth rate of

6.5 percent in 2007, a total trade of US$1.4 billion annually, and aggregate foreign direct investment (FDI) of US$52.4 billion. Levels of economic development range widely, from the cosmopolitan and highly developed Singapore, with a GDP per capita in excess of US$35,000 in 2007, to that of Laos, with a GDP per capita of just US$731 (ASEAN 2009).

Recent demographics and economics do not begin to encompass the total diversity of a region with one of the world's richest arrays of histories, languages, cultures, and religions, including the world's most populous majority-Muslim country (Indonesia), as well as strong currents of Buddhism (Thailand), Christianity, Hinduism (Bali and Java), and numerous local religious traditions. Indonesia alone is comprised of around 14,000 islands, with 600 languages or dialects. Attempts have been made to do justice to higher education in the region (Welch, 2009c), and this chapter builds upon that previous work by focusing on the competitiveness of two higher education systems: those of Malaysia and Vietnam. There are some regional similarities with respect to higher education, but these two systems represent different profiles, especially when measured in terms of the global knowledge system. A study of these two systems underlines the diversity within the region, while allowing a sharper focus than a regional analysis would.

Common to Southeast Asian higher education is a longstanding respect for learning, with some institutions of impressive lineage, such as Hanoi's *Quo Tu Giam* (Temple of Literature), an early site of Confucian higher learning founded by Emperor Ly almost a thousand years ago, pre-dating the ancient universities of Oxford, Cambridge, and Paris.

A second pillar supporting the ubiquitous regional desire to expand the quantity and quality of higher education is a belief in the importance of higher education for social and economic development. A common conviction, from the poorest to the richest countries, and from the smallest to the largest, is that their future is inextricably tied to the vitality of their knowledge and innovation systems (World Bank 2007). For this future to be realized, it is widely agreed that the health of higher education is vital, and thus families often make considerable sacrifices to prepare their children for higher education.

# The Southeast Asian Context

The status and research output of higher education in Malaysia and Vietnam encapsulates much of the diversity that characterizes the region. While Vietnam is one of the poorer nations of the region, Malaysia's GDP per capita level now places it into a middle-income category. Notwithstanding different income levels, however, both systems harbor ambitious plans to extend higher education to larger portions of their populace, who in turn press for more and more higher education institutions (HEIs) for their children. This is for at least two reasons.

First, all regional governments subscribe to the view, advanced by international organizations such as the Organisation for Economic Co-operation and Development (OECD), World Bank, and Asian Development Bank, that higher education is the key production site for the highly skilled personnel who, in a post-Fordist world, are believed to be the foundation of the new knowledge economies. All of the higher education systems of Southeast Asia would subscribe to the following assertion: "The quality of knowledge generated within HEIs, and its availability to the wider economy, is becoming increasingly critical to national [and, one could add, international] competitiveness" (World Bank 2000, 9).

This poses a dilemma for regional governments in terms of international competitiveness. No universities in Indonesia, Malaysia, the Philippines, Thailand, or Vietnam were listed within the top 500 in the 2003 Shanghai Jiao Tong University Institute of Higher Education's *Academic Ranking of World Universities* (MMOHE 2006). That said, each country has cherished higher education icons: Vietnam National University in Vietnam, and both the University of Malaya and *Universiti Sains Malaysia* in Malaysia.

Although both countries' governments, as well as more and more of their populations, see higher education as critical to individual and national progress, there is much that varies between the two states. A key difference arises from the different levels of development in each. Malaysia's GDP per capita in 2002 was US$9,120, compared with that of Vietnam, at US$2,300. Human Development Index (HDI) ranks also differed appreciably, with Malaysia ranked at 59, compared with Vietnam's ranking of 112 (UNDP 2005, 20).

Clearly, these indices place Vietnam well below Malaysia in income but also in terms of HDI ranking. A second trend arises from the severe, but differential, effects of the regional financial crisis of the late 1990s: Malaysia's HDI ranking fell, while Vietnam was the only state in Southeast Asia to emerge from this crisis with its HDI ranking intact (UNDP, 2005). Debt levels are a further distinguishing element: Vietnam's debt, expressed as a percentage of GDP, was a relatively low 3.4 percent, while that of Malaysia was 8.5 percent (UNDP 2005, 27). These elements impose strict limits on public sector efforts in higher education, given that "expenditures on debt servicing and military spending tend to crowd out social expenditures" (UNDP 2005, 26).

A further noteworthy feature is rising levels of public expenditures on education from 1990–2002, although these remain modest compared to developed country averages (UNDP 2005). Substantial differences evident between Malaysia and Vietnam are somewhat misleading, however, since ethnic Chinese and Indian Malaysians, who together make up a third of the Malaysian population, are in practice often excluded from public higher education by ethnic quotas (Tierney 2008; Welch 2008a). Similar cautions apply when analyzing public expenditure on tertiary education—Malaysia spends generously on its public HEIs, from which non-ethnic Malaysians are often in practice excluded.

While comparable Vietnam data is lacking (a familiar problem when analyzing that system), Malaysia's rate of expenditure on education compares well with high HDI countries (UNDP 2007). Once again, however, this is somewhat deceptive, since access, at least to public HEIs, is largely restricted to ethnic Malays (known as *Bumiputras*, or "sons of the soil"). But more than 90 percent of enrollments at private HEIs in Malaysia are from non-*Bumiputras*.

A further constraint arises from net secondary enrollment rates, which remain low relative to more developed economies: 76 percent in Malaysia, and 74 percent for lower secondary school in Vietnam (rates for upper secondary are only about half this) (UNDP 2007; Welch 2007). Many people who do not complete secondary schooling, thereby precluding themselves from higher education, are from the poorest class.

Poverty levels, too constrain further access and equity in higher education in Southeast Asia, varying from a modest 9 percent in Vietnam to 14.1 percent in the Philippines, and falling from 16.5 percent to 6.1 percent in Malaysia just prior to the disastrous regional financial crisis of the late 1990s (World Bank 2007; Asian Development Bank 2008; Welch 2009b). Inequality, particularly between rural and urban groups, remains high, relative to other world regions.

Inequalities of access and participation in education, including higher education, are also common—rates among urban and rural poor lag well behind that of the overall population, in a world region where the state's record of service delivery to the poor has often been weak. Notwithstanding Vietnam's longstanding and impressive commitment to equality, access to higher education is significantly lower outside Hanoi and Ho Chi Minh City (formerly Saigon), while ethnic disparities remain substantial (Welch 2007).

Two recent events also deserve mention. Beginning in 2007 and 2008, sharply increased costs of basic commodities such as food and fuel distorted national budgets supporting the poor, who were hardest hit by swiftly rising prices. Export restrictions were imposed on rice, a key staple, in 2008 (*Sydney Morning Herald*, April 19, 2008). Spiraling food prices, and associated subsidies, did not merely reduce the discretionary income of poor families, thereby putting education out of their reach, but also significantly reduced amounts available from state budgets for education. The second event, the worldwide financial crisis of 2008–2009, is already having an impact on higher education budgets around the world, but it is too early to measure its precise impact in Malaysia and Vietnam. The fact that Malaysia's economy is likely to have contracted by the final quarter of 2008, and Vietnam's previously strong rate of growth slowed significantly, is cause for concern. The regional financial crisis of the late 1990s led to millions being pushed out of education systems, and thousands of Malaysians being brought back from study abroad, to complete their degrees at home (Varghese 2001; Welch 2009b).

Common in much of Southeast Asia, the demographic profile of both Vietnam and Malaysia, relatively young countries with high fertility rates, presents a further challenge, as illustrated in table 11.1. Just addressing these demographic pressures, while also responding to rising aspiration levels for higher education, is a difficult task, even apart from uneven institutional quality.

Flows of foreign direct investment to Malaysia and Vietnam rose significantly after the regional financial crisis of the late 1990s. By 2006, for example, the inflow from *Viet Khieu* (overseas Vietnamese) had risen to US$6 billion, an amount greater than foreign aid or international investment (Welch 2008b). Some of this investment capital supports higher education access and participation, as do some remittances—but the evidence is not systematic (Welch 2009a). FDI flows for Malaysia and

**Table 11.1** Demographic Pressures on Higher Education, Malaysia and Vietnam

|  | Total Population (millions) 1975 | Total Population (millions) 2000 | Annual Population Growth Rate (%) 1975–2000 | Population under 15 (as % of total) 2000 |
|---|---|---|---|---|
| Malaysia | 12.3 | 22.2 | 2.4 | 34.1 |
| Vietnam | 48.0 | 78.1 | 2.0 | 33.4 |

*Source*: UNDP (2002).

Vietnam in 2003 were US$2.474 billion and US$1.450 billion, respectively (UNCTAD 2004).

## Changing Governance Regimes

In a context where no HEIs from Malaysia or Vietnam were listed within the top 500 research universities of the Shanghai index, governance reforms add to institutional pressures, while the lack of income and infrastructure in education also affects regulatory capacity, notably in Vietnam. While regional higher education systems continue to grow, particularly in the private sector, the expansion of both regulatory capacity and transparency has not always matched this.

As the goals set for higher education have been revised internationally against a backdrop of a complex and shifting environment, so too has the governance of higher education (Amaral et al. 2002, 2003; OECD 2003). A key element is the move toward devolution, from a pattern of strong centralization. While governments maintain strong involvement in higher education, some devolution to the institutional level is seen as a means to ensure flexibility and diversity. In both Malaysia and Vietnam, however, public HEIs are said to remain shackled by the state. In Malaysia, traditionally tight controls over many aspects of university life, including the appointment and performance of Vice-Chancellors and matters of academic freedom, have long been criticized, while in Vietnam universities are still expected to promote socialism and maintain official state ideology within a single party state.

At the same time, however, HEIs in both systems are caught in a dilemma that is a consequence of the rapid expansion of the university system: public HEIs, caught between the dual pressures of spiraling enrollments and demand, but lower funding in per-student terms, are increasingly resorting to specific strategies to boost their income, while the expanding private sector responds to unmet demand. Both trends are evident in Malaysia and Vietnam, often at the cost of teaching and research quality, with public sector academics often "poached" to work or moonlight in the private sector (Welch 2007). In Malaysia, public HEIs, corporatized as part of the Seventh Malaysia Plan (1996–2000), have increasingly resorted to developing subsidiaries that run fee-based "executive" programs for those who desire qualifications, but whose grades were insufficient to enter mainstream public HEI courses. These "executive" courses, which often fall outside the Malaysian Qualifications Framework (MQF), are a cause for complaint by dissatisfied students, but are lucrative income sources for the parent institution. They do nothing for the quality or reputation of the parent institution that devised them; indeed, since the programs are often taught by mainstream faculty from the public HEI, they are, in fact, diluting research and teaching quality.

## Quality Assurance

A robust system of quality assurance is key to maintaining the confidence of students, parents, and employers. But while Malaysia and Vietnam each devote specific resources to the regulation of quality in higher education, each has experienced operational difficulties. In Malaysia, the problems are mainly in the private sector, notwithstanding the problems alluded to above in the "executive" programs of public HEIs, which are systematically blurring the borders between public and private. The recently established Malaysian Qualifications Authority, part of the MQF, is the agency vested with this responsibility, yet as was seen above, many programs lie beyond its reach—perhaps no surprise given that there are now more than 500 private colleges. In Vietnam, only modest resources from Vietnam's Ministry of Education and Training (VMOET) can be devoted to the task, and it is not always clear that relevant personnel are trained. At times, too, "poacher" has turned "gamekeeper". The growth of mass higher education systems, and of numbers of HEIs, aggravates quality control difficulties within both the public and private systems.

## Transnational Flows

The challenge of transnational higher education and cross-border programs/institutions makes the regulatory environment even more complex, especially given the rise of numerous bogus "cyber universities" and virtual diploma mills. At the same time, each system suffers from a degree of brain drain: Malaysia because of its longstanding practice of ethnic discrimination, and Vietnam largely because of substantial income disparities with overseas countries. The extent and dimensions of transnational higher education further underline the significant differences between the two, and provide another index of differing levels of international competitiveness. Malaysia now has almost 50,000 international enrollments, a sign of the increasing maturity of its higher education system but also of the attraction of being able to study in English (at least in the private sector), at prices below that offered by the traditional English-language host countries. Vietnam's higher education system cannot begin to match such levels of student inflows.

# The Impact of Corruption

Cronyism is a common complaint in Malaysia, while both corruption and cronyism taint the Vietnamese higher education system. Some of the limits listed above are part of the problem. Transparency International's work underlines the strong correlation between corruption and poverty, with a concentration of poorer states at the bottom of the ranking: "Despite a decade of progress in establishing anti-corruption laws and regulations... much remains to be done before we see meaningful improvements in the lives of the world's poorest citizens" (Transparency International 2006). The Transparency International 2006 Corruption Perceptions Index, a composite index drawing on multiple expert opinion surveys of public sector corruption in 163 countries worldwide, rated countries on a scale from zero to ten (zero indicates high, ten indicates low). Ratings for Malaysia and Vietnam differed significantly. In a context where public sector wages in Vietnam remain poor, and moonlighting is common among public sector HEI faculty in both systems, Malaysia's score (5.0) was double Vietnam's (2.4).

Westcott's (2003) analysis of corruption in Southeast Asia underlines the extent of the problem. Vietnam data showed that nearly one third of Vietnam's public investment expenditure in 1998, "equivalent to 5% of GDP, was lost to fraud and corruption, and the situation hasn't improved since then" (252). This has an evident impact on funding for higher education and ultimately on quality and institutional competitiveness, internationally.

Institutional examples of corruption in higher education relate to both quality and entry criteria. Striking evidence of the extent of the problem emerged in Vietnam in 2001, at certain private HEIs. The first issue involved substantial over-enrollment: the VMOET found Dong Do University to have over-enrolled its quota by 280 percent (4,205 students, rather than 1,500). While the official investigation highlighted institutional malpractice, the problems had been known for years (*Viet Nam News*, June 19, 2002).

The second problem involves entry standards. It was alleged in 2001 that some of Dong Do's leaders had routinely accepted bribes in an effort to boost enrollments and income. Exam entry papers were scored at eight or nine out of ten, at times by unqualified markers, when their real grade was assessed at as low as 0.5. Several dozen students were enrolled despite not being on the selection list. Another 380 had no upper secondary graduation certificates. Overall, 80 percent of Dong Do students had scores lower than that reported by the University Council, while some had marks increased by re-scoring. The investigation also found that the university had not built any facilities, offices, or classrooms in seven years of operation, or invested in faculty development. Facilities were assessed as not meeting the standards of a university (*Viet Nam News*, June 19, 2002, October 7, 2002). In other examples, Ministry officials were involved in the "Asian International University" (AIU) scam, which "enrolled thousands, awarding worthless paper degrees" (*Viet Nam News* June 19, 2002; Le and Ashwill 2004).

While Malaysia is generally freer of corruption, cronyism is an oft-lamented phenomenon, largely arising from the longstanding practice of giving preference to ethnic Malays (*Bumiputras*) over citizens of Chinese or Indian descent. While ethnic quotas purportedly ended in 2003, the longstanding practice of ethnic preference has proved stubbornly resistant to change (*University World News* 2008). In practice, opportunities for non-*Bumiputras* are often still limited, and too few are found above the level of Dean. Promotions are still based on informal ethnic quotas, or cronyism, rather than aptitude, while professional development opportunities are more limited for non-*Bumiputras*.

# The Rise of Private Higher Education

A final dimension of change with critical implications for quality and competitiveness is the swift rise of private higher education. Given such a young population, rising levels of aspiration for higher education, and an already tight budgetary context which is even more challenging in light of the world financial crisis of 2008–2009, neither system has been able to satisfy demand for places in public HEIs. Hence, private higher education has recently grown dramatically, a trend with significant implications for access and equity, given the higher fees generally charged in the private sector. Data from table 11.2 are indicative of the rise of private institutions of higher education in Vietnam and Malaysia, neither of which had a private university sector until recently. This was not due to any affinity of political systems: Vietnam is a socialist polity adapting to the demands of a market economy and recent entry to WTO, while Malaysia represents a capitalist system with a ruling party traditionally bound to ethnic politics,

**Table 11.2**   Numbers and Types of HEIs, Vietnam and Malaysia, 2007

|  | **Public** | | | **Private** | | | |
|---|---|---|---|---|---|---|---|
| Country | Degree | Non-Degree | Subtotal | Degree | Non-degree | Subtotal | Total |
| Malaysia | 18 | 40 | 58 | 22 | 519 | 541 | 599 |
| Vietnam | 201 | — | 201 | 29 | — | 29 | 230 |

*Source*: Asian Development Bank (2008, 45).

but now under significant challenge. Notwithstanding these differences, table 11.2 shows that private higher education has grown vigorously in each country.

Considerable change is evident in each system. Vietnam's current, strikingly ambitious growth targets for higher education entail vigorous growth of private ("People's") HEIs that had already doubled their share from 1996–1997 to 1998–1999 (Welch 2007). Le and Ashwill (2004) report that by 2002–2003, there were 23 private HEIs, enrolling 24,500 students (approximately 12 percent of the total of 200,000 new enrollments). Current government targets are for an astonishing 40 percent to be private by 2020 (Thiep and Hayden 2006; Welch 2008b).

In Malaysia, the rise of private higher education since the mid 1990s is also striking: while no private universities existed then, there are now 18 private universities, with several overseas branch campuses, and private enrollments in higher education that now match those in the public sector (including diploma and certificate levels) (MOHE 2006; Tierney 2008). This dramatic swell in private sector institutions and enrollments, symptomatic of the wider region, has implications not only for access and equity (private HEI fee levels are often much higher than public HEIs), but also for quality. Private HEIs commonly feature lower entry criteria, smaller proportions of qualified faculty (often moonlighting from primary public sector institutions), restricted curricular range, and little if any research.

# Charting Knowledge and Innovation Systems

The dimensions summarized above represent significant challenges to both the quantity and quality of public higher education, notwithstanding substantial differences between the two systems. In particular, the capacity for research and innovation—which is, of course, not limited to higher education—is constrained by financial and demographic factors, as well as the rise of private higher education. These constraints must, however, be set against the shared ambition to extend both the quantity and quality of higher education systems. The discussion below charts levels of achievement for the two systems: it reveals both significant progress and a persistent, if narrowing, gap between the two countries and the global North, in relation to both the density of research and development (R&D) and the relative contribution of HEIs to total R&D.

In terms of what has been called the international knowledge network (Altbach 1998, 2003), it is striking to find the absence of HEIs from either system in the Shanghai index of leading universities. While the issues treated above are part of the explanation, also relevant is the persistently greater density of knowledge indicators in the North. A decade ago, the North, for example, still had approximately ten times the proportion of R&D personnel (scientists and technicians) per capita (3.8 percent) as the South (0.4 percent) and spent about four times the proportion of GDP on R&D (2.0 percent compared to 0.5 percent) (World Bank 2000). Table 11.3 underlines major disparities on a variety of knowledge indices.

While equivalent data were not available for Vietnam, the average difference in researchers between Southeast Asia and the developed world was more than seventeen fold, while even for more-developed Malaysia the difference was twelve fold. In terms of quality of research institutions and university-industry collaboration, however, Malaysia rated just below the developed world average on the former, and somewhat higher on the latter.

Measures of R&D also reveals significant ongoing disparities between Southeast Asia and the developed world, both in terms of total spending and proportion of GDP. Substantial disparities exist, not merely between global North and South, but also within Southeast Asia, although equivalent data for Vietnam were unavailable. World Bank data shows that total R&D spending in Southeast Asia for 2002 was UDS$3.3 billion in Purchasing Power Parity (PPP) terms, which compares to

**Table 11.3**   National Innovation Indices, Malaysia and the Developed World

|  | Average Yrs. of Schooling | Researchers per million | Quality of Scientific Research Institutions | University-Industry Research Collaboration |
|---|---|---|---|---|
| Southeast Asia | 6.6 | 210 | 4.1 | 3.6 |
| Malaysia | 7.9 | 299 | 5.0 | 4.7 |
| Developed Country Average | 9.5 | 3616 | 5.1 | 4.4 |

*Source*: World Bank (2006, 134).

US$645.8 billion for the developed world. Expressed as a percentage of GDP, Southeast Asia spent 0.2 percent on R&D, relative to a developed world average of 2.3 percent (World Bank 2006, 116).

In addition to the disparities between Southeast Asia and the developed world average, the global North registered some 97 percent of all patents in the United States and Europe. When the newly industrializing countries of East Asia (notably Taiwan and South Korea) are added, they accounted for 84 percent of all scientific articles published (World Bank 2000, 69). Recent data from the U.S. Patent Office reveal a continuing wide gap in terms of patents granted. The total number of patents granted to Southeast Asia for 2000–2004 was 140, of which 64 were to Malaysia. This compared with a developed world total of 168,017 (World Bank 2006, 123).

It is important to remember that such indices as the Science Citation Index, Social Sciences Citation Index, and Engineering Index are biased toward English language journals. Acknowledging this additional burden for Malaysia and even more for Vietnam, it is nonetheless illustrative to chart conventional measures of knowledge production (see table 11.4).

Comparative figures help put this into perspective: Australian publications for 1995 totalled 18,088, and for Japan 58,910. Citation counts 1993–1997 were 301,320 and 930,981 respectively.

Figures for knowledge production (papers and citations) for the last decade show significant growth, as depicted in table 11.5, and further underscore substantial differences within Southeast Asia.

**Table 11.4**    Papers and Citations, 1980s and 1990s

| Country | Number of Papers 1981 | Number of Papers 1995 | Number of Citations 1981–85 | Number of Citations 1993–97 |
| --- | --- | --- | --- | --- |
| Malaysia | 229 | 587 | 1,332 | 3,450 |
| Vietnam | 49 | 192 | 203 | 1,657 |

*Source*: World Bank (2000, 125–127).

**Table 11.5**    Papers and Citations, 1998–2008

| Country | Number of Papers 1998–2008 | Number of Citations 1998–2008 | Citations per paper 1998–2008 |
| --- | --- | --- | --- |
| Malaysia | 14,782 | 60,856 | 4.12 |
| Vietnam | 5,207 | 31,959 | 6.14 |

*Source*: Thomson Science Watch (2009).

While the wealthier and more developed system of Malaysia was responsible for almost three times the total papers produced by Vietnam, and almost twice the total citations, Malaysia's rate of citations per paper is surprisingly weak, relative to Vietnam (and other Southeast Asian systems). Science Watch data shows that Malaysia's citations per paper, for example, rose from 1.66 over the period 1998–2002, to 2.10 for the years 2003–2007, compared with Vietnamese citation rates of 2.32 and 3.50 for the same periods. Taken together, Tables 11.4 and 11.5 reveal that while each system expanded its innovation indices significantly, Vietnam's progress seems more striking.

Once again, comparative figures underline the difference between the two systems considered here and the more developed nations of the Asia Pacific: over the same period, Japan was responsible for 806,008 papers and 7,397,444 citations (citation count of 9.18), and Australia 271,311 papers and 2,870,552 citations (citation count 10.58). The trend, however, is toward a significant narrowing of the performance differential between Malaysia and Vietnam, in terms of total papers produced, and the equivalent totals for both Australia and Japan, over the last decade or more. In each case, and more notably by Vietnam, the gap, while still significant, has narrowed appreciably.

Other recent data show that regional higher education contributes much less to total R&D performance than among developed nations. The proportion the higher education sector contributes toward total R&D performance in Southeast Asia and Malaysia is barely half the average of developed nations. The respective figures are 15.7 percent, 14.4 percent, and 27 percent (World Bank 2006, 120).

Overall, while significant disparities exist between rates and density of knowledge production between Southeast Asia and the developed world, two points should be borne in mind. Firstly, differences exist, notwithstanding traditions of great respect for education and the teacher. For example, Vietnam's Temple of Literature—refurbished some years ago by American Express—contains the scholar-priests of many centuries ago (Welch 2008b). Secondly, the fact that the gap is narrowing, at least on certain measures, is a sign of the commitment of national governments to stimulate knowledge and innovation. If continued, a significant further narrowing of the gap should soon be evident.

# Conclusion

Despite possessing the highest annual GDP per capita growth rate of any world region in recent decades (World Bank 2006; Welch 2008b), as well as high aspirations for higher education among wider and wider proportions of the population and a high commitment to learning, several factors limit the capacity to increase the competitiveness of knowledge and innovation systems in Malaysia and Vietnam. Existing levels of infrastructure, a young demographic profile, the rise of private higher education, limited finances, transparency issues, and limited regulatory capacity represent key constraints on knowledge creation, notably in Vietnam. Persistent controls by the government also limit the freedom and independence necessary for knowledge to flourish. The above analysis also underlines the importance of within-group differences, with countries such as Malaysia being generally better placed to extend their lead. However, Vietnam's rate of progress, with respect to measures of knowledge and innovation, is impressive.

The rise of global English (Crystal 1997; Wilson et al. 1998), in particular as the language of science, represents a further challenge to regional knowledge and innovation systems, to which Malaysia, where English is more widely spoken, is better placed to respond—at least in the private sector, which is less bound by the prohibition of teaching in languages other than *Bahasa Malaya* at the undergraduate level.

The capacity of transnational education—including the critical issues of brain drain and deployment of diaspora to contribute to the competitiveness of national innovation systems—is another discriminating factor (Welch 2008c). It connects, too, to the rise of global English, since academic talent tends to be attracted to major English language systems. While it is clear that major systems such as the United States, United Kingdom, and Australia all continue to benefit by attracting highly skilled students and academics, who then contribute to knowledge and innovation, this is much less the case in Southeast Asia, which continues to lose talent. However, there are substantial differences. Of the two systems, Malaysia, with some 50,000 international students enrolled (primarily in its private system, approximately one third from China), has clearly benefited from its strategy to position itself as an "Eduhub" within the region. While China and Indonesia are examples of major sources for Malaysia's higher education system, Bangladesh, the third largest source country, reflects a pattern of entry from Islamic countries (Welch 2009b). While precise measures of the contribution to knowledge and innovation indices are difficult, Malaysia's policy regarding the quality of its international students is designed to assist in achieving excellence, as the Minister of Education confirmed recently on a visit to the Middle East: "Governments will only sponsor the best students and we need good students to help improve the rankings of our universities. Whatever we do, we can't run away from achieving excellence" (Bernama 2009).

The planned development of Apex ("world-class") universities in both Malaysia and Vietnam represents another attempt to boost knowledge and innovation indices and international rankings. It is too early to assess the impact of such moves, but Malaysia's recent naming of *Universiti Sains Malaysia* as its first Apex institution, with substantial additional funding and an expectation that it will soon number within the top 100 universities, represents the most substantial initiative. Vietnam, too, has announced plans to develop Apex universities, but in a context where resource levels, infrastructure and quality of teaching and research staff are different than Malaysia. Its ambition is less likely to be matched by achievement, notwithstanding the recent impressive progress on indices of innovation.

The explosive growth of private higher education in Malaysia and Vietnam presents a challenge to efforts by each to enhance the competitiveness of their knowledge and innovation systems or boost university rankings, since most private HEIs lack a research function; indeed, many would rather be classed as demand-absorbing. The ongoing rise of private higher education (Altbach 1999) poses a further challenge to quality by infringing equality of access:

> There is another important downside to private financing—it may preclude the enrollment of deserving students who do not have the ability to pay, and often evokes resentment among students who do. Means-tested scholarship and loan programs are one possible approach to addressing this problem, but they have proven difficult to administer due to the difficulty of assessing ability to pay, sometimes exorbitant administrative costs, corruption and high rates of default. (World Bank 2000, 57)

If quality implies equality (Welch 2000), but private higher education in Malaysia and Vietnam generally charges significantly higher fees than public HEIs, the effect is to effectively preclude able poor students from access—something that represents a dilution of quality, rather than its enhancement. This, too, provides no support to national drives to boost university rankings, and the overall competitiveness of higher education systems.

This problem assumes greater proportions, given the recent pressure on public-sector HEIs to increase their income, and diversify sources of funding. In effect, many public HEIs in both Malaysia and Vietnam now often act more like private providers. Evidence for this is seen in two trends. First, both states have introduced fee-based "executive" or "extension" courses, which are offered at night or on weekends, and with lower entry criteria. These are usually taught by selected mainstream faculty, who may profit handsomely from the additional income (as does the institution).

But such courses often do not qualify graduates for employment in the public sector, and at least in Malaysia, are often cause for complaint by dissatisfied students (Welch 2009b). Such programs do little to enhance the quality of teaching or research, and probably weaken the research mission of universities.

Second, both systems, but particularly Vietnam, require better and more comprehensive regulation and quality assurance. This is a consequence of the dramatic rise of private higher education, including some international private providers that are no more than shopfronts or (cyber) diploma mills. Corruption in Vietnam and cronyism in Malaysia limit the capacity for enhanced competitiveness. Neither contributes to enhanced quality or rising knowledge and innovation indices, while in Malaysia the persistence of ethnic discrimination in access to, and advancement in, higher education, despite its formal abandonment in 2003, artificially limits the talent pool, forcing many Chinese and Indian Malaysians into the private sector, or abroad (from which some do not return).

The above challenges to raising the competitiveness of higher education in Vietnam and Malaysia, including efforts to boost university rankings, occur despite substantial recent progress, the history of widespread respect for learning, and the significant commitment by students, families, and governments. Measured by the knowledge indices above, both systems have made great strides. Malaysia has projected an international presence, and has wider ambitions, while Vietnam's achievements are striking, if of a lower base. For both countries, what difference Apex universities will make remains to be seen. At the same time, while there are significant differences between the wealthier and more developed Malaysian system, and the less well-established system of Vietnam, the fact that no university from either system is yet listed in the Shanghai index means that both systems are likely to remain part of the knowledge periphery for the shorter term, while continuing to close the gap on the more developed systems.

# References

Altbach, Philip G. 1998. "Gigantic Peripheries: India and China in the World Knowledge System." In *Comparative Higher Education. Knowledge, the University and Development*, ed. P. G. Altbach. Greenwich, CT: Ablex.
———. ed. 1999. *Private Prometheus: Private Higher Education and Development in the 21st Century*. London: Greenwood Press.
———. 2003. "Centres and Peripheries in the Academic Profession: the Special Challenges of Developing Countries." In *The Decline of the Guru: The Academic*

*Profession in Middle and Lower Income Countries*, ed. P. G. Altbach. New York: Palgrave Macmillan.

Amaral, Alberto, Glen A. Jones, and Berit Karseth, eds. 2003. *Governing Higher Education: National Perspectives on Institutional Governance.* Dordrecht, The Netherlands: Kluwer.

Amaral, Alberto, Lynn V. Meek, and Ingvild Marheim Larsen, eds. 2002. *The Higher Education Managerial Revolution?* Dordrecht, The Netherlands: Kluwer.

Association of Southeast Asian Nations. 2009. *Basic ASEAN Indicators.* Jakarta: Association of Southeast Asian Nations. http://www.aseansec.org.

Asian Development Bank. 2008. Vol. 38 of *Key Indicators. Inequality in Asia.* Manila: Asian Development Bank.

Bernama. 2009. "Malaysia to Boost Muslim Progress through Education." Kuala Lumpur: Bernama. http://www.bernama.com.

Crystal, David. 1997. *English as a Global Language.* Cambridge, MA/New York: Cambridge University Press.

Le, Ngoc Minh, and Mark Ashwill. 2004. "A Look at Nonpublic Higher Education in Vietnam." *International Higher Education* 36 (6): 16–17.

Malaysia's Ministry of Higher Education (MMOHE). 2006. *Report by the Committee to Study, Review and Make Recommendations Concerning the Development and Direction of Higher Education in Malaysia.* Putrajaya, Malaysia: MMOHE.

Organisation for Economic Co-operation and Development. 2003. *Education Policy Analysis 2003.* Paris: OECD.

*Sydney Morning Herald.* 2008. "We Fill Our Tanks while They Can't Fill Their Stomachs." *Sydney Morning Herald,* 19 April. http://www.smh.com.au.

Thiep, Lam Quang, and Martin Hayden. 2006." A 2020 Vision for Higher Education in Vietnam." *International Higher Education* 44 (Summer): 11–12.

Thomson Science Watch. 2009. "Country Profiles." Philadelphia: Thomson Science Watch. http://sciencewatch.com.

Tierney, William. 2008. "Forms of Privatization: Globalization and the Changing Nature of Tertiary Education." Unpublished manuscript.

Transparency International. 2006. *2006 Corruption Perceptions Index.* Berlin: Transparency International. http://www.transparency.org.

United Nations Conference on Trade and Development. 2004. *World Investment Report 2004. The Shift Towards Services.* New York and Geneva: United Nations Conference on Trade and Development.

United Nations Development Programme. 2002. *Human Development Report 2002 Deepening Democracy in a Fragmented World.* New York: UNDP.

———. 2005. *Human Development Report for South East Asia.* New York: UNDP.

———. 2007. *Human Development Indicators 2007/2008.* New York: UNDP.

*University World News.* 2008. "Malaysia: Inter-ethnic Tensions Touch Universities." London: *University World News.* http://www.universityworldnews.com.

Varghese, N. V. 2001. "Impact of the Economic Crisis on Higher Education in East Asia: An Overview." In *Impact of the Economic Crisis on Higher Education in East Asia: Country Experiences,* ed. N. V. Varghese. Paris: United Nations Educational, Scientific and Cultural Organisation.

*Viet Nam News.* 2002. "Officials Fall in School Scandal." *Viet Nam News,* June 19.
http://vietnamnews.vnagency.com.vn.

———. 2002. "Police Grill Professor Over Dong Do University Scandal." *Viet
Nam News,* October 7. http://vietnamnews.vnagency.com.vn.

Welch, Anthony R. 2007. "Ho Chi Minh Meets the Market. Public and Private
Higher Education in Vietnam." *International Education Journal* 8 (3): 35–56.

———. 2008a. "Imagining Islam: Funding and Governance of Islamic Higher
Education in Indonesia and Malaysia." In *Positioning the University in the
Globalized World: Changing Governance and Coping Strategies in Asia.* Hong
Kong: Hong Kong University.

———. 2008b. "Internationalisation of Vietnamese Higher Education: Retrospect
and Prospect." In *The Modernisation of Vietnamese Higher Education,* ed.
G. Harman, M. Hayden, and T. Pham. Amsterdam: Springer.

———. 2008c. "Myths and Modes of Mobility. The Changing Face of Academic
Mobility in the Global Era." In *Going Global. Academic Mobility in Higher
Education,* ed. M. B. Dervin. Cambridge, MA: Cambridge Scholars Press.

———. 2009a. "Financing Access and Equity in SE Asian Higher Education: State
Capacity, Privatisation, and Transparency." In *Financing Higher Education:
Access and Equity,* ed. J. Knight. Rotterdam: Sense Publishers.

———. 2009b. *Higher Education in Southeast Asia. Blurring Borders, Changing
Balance.* London: Routledge.

———. 2009c. "Measuring Up? The Competitive Position of Southeast Asian
Higher Education." In *Critical Issues on Quality Assurance and University
Rankings in Higher Education in the Asia Pacific,* ed. S. Kaur and W. Tierney.

Westcott, Clay. 2003. "Combating Corruption in Southeast Asia." In *Fighting
Corruption in Asia: Causes, Effects and Remedies,* ed. J. B. Kidd and F. J. Richter.
Singapore: World Scientific Press.

Wilson, Mary, Adnan Qayyam, and Roger Boshier. 1998. "World Wide America:
Manufacturing Web Information." *Distance Education* 19 (1): 109–141.

World Bank. 2000. *Higher Education in Developing Countries: Peril and Promise,
Task Force on Higher Education and Society.* Washington, DC: World Bank.

———. 2006. *An East Asian Renaissance. Ideas for Economic Growth.* Washington,
DC: World Bank.

———. 2007. *Malaysia and the Knowledge Economy: Building a World Class Higher
Education System.* Washington, DC: World Bank.

# Chapter 12

# China's Drive for World-Class Universities

*Kathryn Mohrman and Yingjie Wang*

In the past 15 years, the People's Republic of China has pursued two higher education policies simultaneously—massification on one hand, and development of world-class universities on the other. These twin strategies have produced the largest tertiary system in the world, with more than 27 million students in 2007 and rapid enhancement of the nation's top universities.

This chapter focuses on a policy supported by the government in which the nation's top universities receive special funding to become internationally competitive through a series of elitist programs. Campus-level case studies analyze the impact of governmental investment and provide new analysis on the extent to which the policy goal is being achieved.

## Context

More than a decade ago, China's president declared that the nation's modernization requires first-quality universities. The main policy vehicle to achieve this goal is the 985 Project (named for its May 2008 announcement date), directing substantial funding to 39 leading institutions. In addition, the national government created the 211 Project in 1995, designed to develop 100 top universities for the twenty-first century (hence the name).

China has also established a National Science Foundation, National Social Science Foundation, and other programs for research funding, much of which flows to 985 and 211 universities.

These programs developed in a cultural context of elitism. Historically, Chinese universities admitted three to 5 percent of the age cohort based on examination scores. After the Cultural Revolution of 1966–1976 ended, universities grew, while thousands of students were sent to study abroad to bring the latest academic knowledge to China.

Over the past two centuries, the cultural context also involves China's self-perception as a weak country. While China was never formally a colony of a Western power, large portions of the country were controlled by European nations, the United States, Russia, and Japan through military and economic means. At the same time, China considers itself one of the greatest civilizations in the world. Some Chinese commentators argue that western nations want to prevent China from resuming its rightful place in the international sphere. Thus, the drive to create world-class universities is both a rational and an emotional policy response to the cultural context.

Today, China emphasizes education as essential for economic and social development. At the national level, the Ministry of Education has allocated substantial sums for the 985 Project. The government sets broad guidelines for the project and the chosen universities make proposals for multi-year funding within stated priorities.

In this chapter, we analyze China's public investment in higher education through three case study institutions—Sichuan University (SCU), Tianjin University (TJU), and Beijing Normal University (BNU), all participants in the 985 Project. In addition, we provide data from two leading Chinese institutions, Peking University (PKU) and Tsinghua University (THU). Three U.S. universities are used for comparison purposes—Massachusetts Institute of Technology (MIT), University of California, Berkeley (UCB), and University of Michigan, Ann Arbor (UM).

## Projects 211 and 985

The 211 Project, which began in 1995, was the first effort by the Chinese government to strengthen higher education through support for key disciplinary areas, an improved internet system, and overall institutional capacity. The first phase (1996–2000) represented the largest investment in higher education since 1949.

In the second phase (2001–2005), the central government allocated six billion Renminbi (RMB, approximately US$700 million) while an additional 12.4 billion RMB (US$1.5 billion) came from other ministries, provincial governments and matching funds raised by the universities themselves. In the current phase (2006–2010) these institutions continue to receive support at approximately the same level. The 211 Project set the precedent of special funding for a small number of universities, an elitist policy pursued further by the 985 Project.

Project 985 is designed to promote innovation and creativity as well as international competitiveness. The major areas supported by the project are management reform, faculty development, creation of research bases and centers, infrastructure upgrades to support instruction and research, and expanded international cooperation. The first phase of the project (1999–2003) provided a total grant of 14.2 billion RMB (US$1.7 billion) to 39 universities. The second phase of the project (2004–2008), the focus of this chapter, totals 19.1 billion RMB (US$2.3 billion) in support.

# The Case Study Universities

Sichuan University was established in 1896 under an imperial edict. For decades, it has been a large, multi-disciplinary university with departments ranging from literature to agriculture to engineering. Through mergers in 1994 and 2000, the institution added significant strength in science, technology, and medicine. Today, SCU has both the widest coverage of disciplines and also the largest scale of operation in Western China.

Tianjin University was established in 1895, under the name of Peiyang University, as the first modern university in China. In its early years, it followed the U.S. model and emphasized Western learning to train personnel for the nation's industrialization. In 1984 Tianjin University developed plans to become a world-class, progressive, and multidisciplinary research university with a focus on engineering integrated with science, liberal arts, business, and law. One of its missions is "bringing knowledge to industry for social development" (TJU 2009).

Beijing Normal University was established in 1902 as the Faculty of Education of the Imperial University of Peking. Since its establishment, BNU has been considered the top teacher education institution in China. Its motto, "Learn, so as to instruct others, Act, to serve as example to all," reflects its mission for the past century. BNU's current vision is: "We pursue international prestige, aiming to be a comprehensive and research-oriented university with the specialization in teacher education and educational science" (BNU 2009).

# Case Study Analysis

These three institutions, along with the comparison universities listed above, provide an opportunity to analyze the impact of the 985 Project at the institutional level. We collected statistics from public documents as well as campus sources to analyze the progress of the three case study universities toward the goal of international competitiveness.

## Enrollment

Sichuan University is by far the largest of the eight universities examined in this chapter. At nearly 60,000 students in 2007, it is almost twice the size of Tianjin or Tsinghua. The University of California, Berkeley and the University of Michigan, Ann Arbor, are similar large public institutions, each had fewer than 40,000 students in 2007. The smallest institution in the sample is MIT with 8,300 students in 2007.

While the national policy for higher education in China continues to seek growth, the government's emphasis on expansion has shifted from national universities to provincial, vocational, and adult education institutions. With the exception of Sichuan, where the enrollment increased by 42 percent between 2003 and 2007, the Chinese universities in this study demonstrate this new emphasis. TJU grew only 3.2 percent and Tsinghua actually became smaller.

Particularly in national universities, undergraduate enrollment remained relatively steady while the number of graduate students has increased for several reasons. First, top universities have sought to emphasize graduate education as more appropriate for their research missions. Second, graduate study in most universities is free for students, while undergraduate programs require tuition payments. Finally, with the dramatic increase in enrollments in recent years, millions of bachelor's degree graduates have had difficulty finding employment. If more college graduates seek advanced degrees, their entry into the labor market will be delayed and their employment chances presumably will improve.

Emphasis on graduate education varies, however, by institution. About 60 percent of all MIT students are in graduate or first-professional degree programs, while comparable enrollments at BNU, PKU, and THU represent about half of the student body. Few would dispute the research orientation of Berkeley or Michigan, yet these universities also have a public mission of providing education to large numbers of citizens in their respective states; hence, they have large undergraduate enrollments. SCU shares this goal, with about 70 percent undergraduate students.

## Faculty and Staff

Table 12.1 shows the composition of the employee base for the institutions in this study. It is interesting to note that in 2007, the number of non-faculty researchers equals the number of faculty at Tsinghua and far exceeds the number of faculty at MIT.

In China there are several reasons for the rise in professorships. The rapid increase in enrollment has required an increase in faculty, although the growth in teachers has not matched growth in student numbers. Also, as these institutions ramp up their research effort, they need additional faculty to conduct research. Most of the case study universities have increased the proportion of faculty in their total workforce, with Beijing Normal faculty members exceeding 50 percent of all employees in 2007 and Sichuan not far behind. Michigan, too, increased the number of professors between 2003 and 2007.

While the universities in this study are all research-intensive institutions, it is interesting to note that they have improved (i.e., reduced) their student-faculty ratios, in some cases quite dramatically, between 2003 and 2007—Tsinghua and Michigan in particular—through increases in faculty numbers. In 2007 the lowest student-faculty ratio was 6.1 students

**Table 12.1**    Number of University Employees by Major Activity

|  | 2003 | | | | 2007 | | | |
|---|---|---|---|---|---|---|---|---|
|  | Faculty | Researchers* | Staff** | Total | Faculty | Researchers* | Staff** | Total |
| SCU | 3,128 | 1,042 | 3,967 | 8,137 | 3,946 | 451 | 3,674 | 8,071 |
| TJU | 1,722 | 342 | 2,516 | 4,580 | 1,969 | 331 | 1,961 | 4,261 |
| BNU | 1,225 | — | 1,490 | 2,715 | 1,660 | 70 | 1,421 | 3,151 |
| PKU | 2,409 | 864 | 4,250 | 7,523 | 2,926 | 1,381 | 5,168 | 9,475 |
| THU | 2,022 | 2,157 | 3,937 | 8,116 | 2,580 | 2,033 | 6,178 | 10,791 |
| MIT | 1,253 | — | 9,578 | 10,831 | 987 | 3,720 | 9,019 | 13,726 |
| UCB | 1,901 | 1,356 | 12,831 | 16,088 | 2,139 | 852 | 13,744 | 16,735 |
| UM | 4,313 | — | 17,209 | 21,522 | 6,374 | — | 16,802 | 23,176 |

*Notes*:

* Researchers are non-faculty employees whose primary responsibility is research, not teaching (not all institutions report this as a separate category).

** Staff are employees who are not faculty and not researchers.

*Sources*: CMOE (2008); NCES (2009).

per professor at Michigan and the highest was 16.3 at Berkeley. The major exception to this trend is Sichuan, where a 26 percent increase in faculty could not keep pace with 42 percent enrollment growth.

## University Finances

Table 12.2 shows that university expenditures for all purposes have grown dramatically on the eight campuses. The reasons for increasing expenditures in China are many: increasing numbers of undergraduate students mean more tuition fees, research grants have risen, universities are becoming more entrepreneurial, alumni are being asked to make contributions, and the 985 Project is adding significant resources to China's top universities.

Comparisons are difficult since the two countries have quite different economic structures. To reduce such variability, the World Bank uses Purchasing Power Parity (PPP) to enable meaningful comparisons across nations. PPP investigates the cost of a market basket of similar goods and services in different countries to determine the real buying power of each local currency. The most recent World Bank comparison study (2008) pegs each national situation to one U.S. dollar equaling 1.00. In this system, Chinese RMB should be divided by 3.45 in order to equal one U.S. dollar—thus, it takes 3.45 Chinese RMB to buy the same goods as one U.S. dollar.

Figure 12.1 displays the statistics from table 12.2 converted using PPP. Not surprisingly, the data show that Chinese universities are less wealthy than their U.S. peers. Sichuan, with more students than Michigan, had a

**Table 12.2**   Total University Expenditures, 2003, 2007

|      | 2003          | 2007          | % Change |
|------|---------------|---------------|----------|
| SCU  | 937,028,200   | 1,703,188,300 | 81.8     |
| TJU  | 849,415,100   | 1,101,060,100 | 29.6     |
| BNU  | 806,808,400   | 1,197,223,900 | 48.4     |
| PKU  | 2,127,509,400 | 3,753,312,500 | 76.4     |
| THU  | 3,100,112,500 | 4,437,911,900 | 43.2     |
| MIT  | 1,686,573,000 | 2,207,621,000 | 30.9     |
| UCB  | 1,416,147,000 | 1,685,527,000 | 19.0     |
| UM   | 3,556,243,000 | 4,433,602,000 | 24.7     |

*Note*: Statistics are given in Chinese RMB for Chinese universities and U.S. dollars for American universities (not adjusted for inflation).

*Sources*: CMOE (2008), NCES (2009), MIT (2008), UCB (2008), and UM (2008).

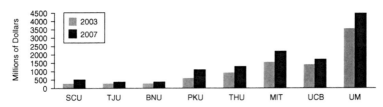

**Figure 12.1**  University Expenditures in US$ Using PPP, 2003 and 2007

*Sources*: MIT (2007), UCB (2007), UM (2007), and CMOE (2008).

**Table 12.3**  Average Annual Grant from 985 Project as Percent of Total University Expenditures, 2007

|  | 985 Project grant Phase 2 (2004–2008) | Average annual grant as percent of total univ expenditures |
|---|---|---|
| SCU | 400,000,000 | 5.9 |
| TJU | 421,000,000 | 9.6 |
| BNU | 600,000,000 | 12.5 |
| PKU | 2,408,000,000 | 16.0 |
| THU | 3,591,000,000 | 20.2 |

*Note*: Amounts provided in Chinese RMB

*Sources*: CMOE (2008) and institutional data.

lower total expenditure for 2007 and Tsinghua is nearing Berkeley, even though it has many fewer students.

One way to judge the impact of the 985 Project on Chinese university budgets is to compare its awards to individual universities with the size of the institutions' resources (table 12.3). Funds from the 985 Project are allocated in four-year grants; although the actual disbursement is not made in equal amounts each year, for purposes of this comparison, the amount for 2004–2008 is divided into four parts. The impact of the project in financial terms ranges from about 6 percent at Sichuan to more than 20 percent at Tsinghua. With a significant proportion of the annual budget coming from the 985 Project, Tsinghua is able to support its faculty and its research enterprise at higher levels than the other universities in this study.

## Research Funding

China has put significant emphasis on university-based research, both for contributions to economic development and also for international competitiveness. The 985 Project cannot fund research directly; however, the money has been used extensively for equipment, laboratories, and other research-related purposes.

The Chinese universities in this study have increased research funding dramatically between 2003 and 2007. The growth is especially impressive because, until the 1980s, research was conducted almost exclusively in separate research institutes (e.g., the Chinese Academy of Sciences) while university professors were focused on the training of young scholars. Thus, the tradition of university research is quite young, leading the government to see the need for major support.

Figure 12.2 takes research and development (R&D) funds from external sources in both Chinese and U.S. universities and equates them, using PPP. As one might expect, the three U.S. institutions all spent more than their Chinese counterparts. More surprisingly, in 2007 Tsinghua spent about two-thirds as much as did MIT, with a faculty/research staff of approximately the same size, an impressive showing for a university in a rapidly developing country.

The universities in question vary significantly in size, so a per-capita statistic is a useful way to compare their commitment to research. Since several of the universities in this study have significant numbers of non-faculty researchers, table 12.4 calculates R&D expenditures per professor and R&D expenditures per professor plus researcher. The table shows that the growth in per-person funding between 2003 and 2007 has been fastest at Sichuan University, reflecting a substantial increase in total research expenditures. In just four years, both Peking and Tsinghua increased their per-person R&D expenditures by an impressive 50 percent. Surprisingly, Berkeley and Michigan show negative per-capita changes per professor, reflecting relatively modest growth in research funding between 2003 and 2007 at the same time as they had significant increases in the number of faculty members.

These statistics suggest that Peking and Tsinghua are far ahead of the other Chinese universities in this study. While the others are increasing

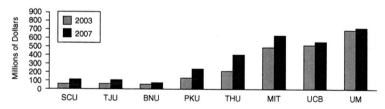

**Figure 12.2**   Research Expenditures in US$ Using PPP, 2003 and 2007

*Sources*: CMOE (2008) and NSF (2008).

**Table 12.4** R&D Expenditures per Professor and per Professor plus Researcher, 2003 and 2007

| | 2003 | | 2007 | | | |
|---|---|---|---|---|---|---|
| | Expend per prof. | Expend per prof. plus researcher | Expend per prof. | Expend per prof. plus researcher | Percent change professor | Percent change prof. & researcher |
| SCU | 54,812 | 41,116 | 96,701 | 86,783 | 76.4 | 111.1 |
| TJU | 111,023 | 92,627 | 146,836 | 125,704 | 32.3 | 35.7 |
| BNU | 89,294 | — | 108,920 | 104,513 | 22.0 | — |
| PKU | 180,731 | 133,022 | 271,394 | 184,374 | 50.2 | 38.6 |
| THU | 353,234 | 170,911 | 536,300 | 299,947 | 51.8 | 75.5 |
| MIT | 387,681 | — | 622,444 | 130,519 | 60.6 | — |
| UCB | 266,800 | 155,722 | 258,235 | 184,676 | -3.2 | 18.6 |
| UM | 180,861 | — | 126,880 | — | -29.8 | — |

*Note:* All expenditures are given in U.S. dollars; Chinese statistics are converted using PPP.

*Sources:* CMOE (2008), NSF (2008), and NCES (2009).

their research investment, and dramatically so, they are not in the same league as PKU and THU. Perhaps only these two well-funded universities stand a chance to be truly competitive in the international arena for the foreseeable future.

## Publications

What are universities "buying" with their R&D expenditures? In general terms, they are investing in new and applied knowledge, and the most common way of measuring that investment is through scholarly publications. Indeed, most ranking systems and league tables use publications, especially those in highly regarded and peer-reviewed journals, as a proxy for quality of research.

As table 12.5 shows, the Chinese universities in this study have increased their scholarly productivity substantially, in two cases (TJU and BNU) by

**Table 12.5**    Total Publications by Faculty and Staff, 2003–2007

|  | 2003 | 2007 | Percent Change |
|---|---|---|---|
| **SCU** | | | |
| Natural sciences | 1,532 | 1,234 | −19.5 |
| Social sciences and humanities | 3,946 | 6,395 | 62.1 |
| Total publications | 5,478 | 7,629 | 39.3 |
| **TJU** | | | |
| Natural sciences | 3,538 | 6,719 | 89.9 |
| Social sciences and humanities | 134 | 158 | 17.9 |
| Total publications | 3,672 | 6,877 | 87.3 |
| **BNU** | | | |
| Natural sciences | 1,048 | 1,623 | 54.9 |
| Social sciences and humanities | 1,360 | 2,743 | 101.7 |
| Total publications | 2,408 | 4,366 | 81.3 |
| **PKU** | | | |
| Natural sciences | 3,061 | 3,507 | 14.6 |
| Social sciences and humanities | 2,438 | 2,286 | −6.2 |
| Total publications | 5,499 | 5,793 | 5.3 |
| **THU** | | | |
| Natural sciences | 6,494 | 9,617 | 48.1 |
| Social sciences and humanities | 1,386 | 1,899 | 37.0 |
| Total publications | 7,880 | 11,516 | 46.1 |

*Source*: CMOE (2008).

more than 80 percent between 2003 and 2007. Expectations for faculty members have risen as Chinese academics acknowledge the importance of publications for external recognition. Professors in both countries now realize that publications are essential for salary increases and promotion. Many incentives, therefore, all work in the same direction—more academic publications by faculty and staff.

But not all publications are equal. Thomson Reuters, an international data firm, has created several groupings of quality academic publications, the articles in which are listed in Science Citation Index, Social Sciences Citation Index, and Arts and Humanities Citation Index. Most of the journals in these indices are published in English. If professors have articles accepted in one of the journals included in the indices, they have "cited publications" that are considered to have met high standards of peer review. Thus, it is more prestigious to produce cited articles than other publications, a challenge for many international scholars whose first language is not English. Only PKU has the majority of its total publications in cited journals; the other Chinese universities range from 21 to 44 percent.

Table 12.6 adjusts the publications figures on a per-capita basis to account for the differences in size of the universities. Not surprisingly, MIT has the largest number of cited articles per professor, but when its large researcher corps is figured in, its per-person production rate falls dramatically. In 2007 both Peking and Tsinghua equal MIT in cited publications per professor and researcher. Berkeley is by far the most productive of all the institutions in this study in 2007.

R&D expenditures per publication (figure 12.3) demonstrate how much it costs to produce a high quality article. It is not surprising that

**Table 12.6** Cited Articles per Professor and per Professor and Researcher, 2007

|  | Number of cited articles | Number of professors | Number of Prof. & researchers | Cited articles per professor | Cited articles per prof. & researcher |
|---|---|---|---|---|---|
| SCU | 1696 | 3946 | 4397 | 0.4 | 0.4 |
| TJU | 1350 | 1969 | 2300 | 0.7 | 0.6 |
| BNU | 906 | 1660 | 1730 | 0.5 | 0.5 |
| PKU | 4516 | 2026 | 3407 | 2.2 | 1.3 |
| THU | 4428 | 2580 | 4613 | 1.7 | 1.0 |
| MIT | 3763 | 987 | 4707 | 3.8 | 0.8 |
| UCB | 5109 | 2139 | 2991 | 2.4 | 1.7 |
| UM | 5803 | 6374 |  | 0.9 |  |

*Sources*: CMOE (2008) and ISI (2009).

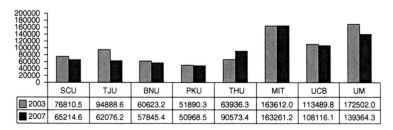

**Figure 12.3**    Research Expenditures per Cited Article in US$ Using PPP, 2003 and 2007

*Sources*: NSF (2008) and ISI (2009).

the cost of doing research, as measured by publications, is higher in the United States than in China. It is interesting to note, however, that almost all the studied universities reduced their cost per article between 2003 and 2007, both because the number of publications increased rapidly, and also because professors are becoming more successful in publishing in international journals. In addition, universities are increasing their efficiency in managing the R&D effort.

In contrast to the general tendency, MIT maintained the same level of expenditure per article between 2003 and 2007 and at Tsinghua the cost per article rose an astonishing 41.7 percent. Earlier in this chapter, Tsinghua was praised for its research output, compared to MIT, Berkeley and Michigan, but Figure 12.3 shows that Tsinghua's output is very expensive—and increasingly so. Of course, Tsinghua may be investing now for research results that will be published years into the future. In addition, the expense may be related in part to the substantial support provided by the 985 Project.

## University Rankings

Another way to assess competitiveness is through international comparisons. One of the most frequently used ranking systems is the *Academic Ranking of World Universities* (ARWU), produced annually by Shanghai Jiaotong University Institute of Higher Education (SJTUIHE 2008). Table 12.7 displays several of the criteria used in ARWU, largely focused on research and publications. In 2003, SCU, TJU, and BNU were not included among the top 500 worldwide, although PKU and THU both appeared in the alphabetical listing of institutions ranking between 200 and 300. The U.S. comparison universities, on the other hand, were highly

**Table 12.7** Academic Ranking of World Universities (Shanghai Index)

| | Shanghai ranking | | Cited articles | | Cited researchers | | Per capita performance | | Total score | |
|---|---|---|---|---|---|---|---|---|---|---|
| | 2003 | 2008 | 2003 | 2008 | 2003 | 2008 | 2003 | 2008 | 2003 | 2008 |
| SCU | Not incl. | 402–503 | | 0 | 0 | 0 | | 40.1 | | 11.5 |
| TJU | Not incl. | 402–503 | | 3.4 | 0 | 0 | | 34.5 | | 13.3 |
| BNU | Not incl. | not incl. | | | | | | | | |
| PKU | 251–300 | 201–302 | 49.9 | 11.7 | 0 | 0 | 5.9 | 54.6 | | 16.3 |
| THU | 201–250 | 201–302 | 54 | 9.4 | 0 | 0 | 10.4 | 56.3 | | 18.1 |
| MIT | 6 | 5 | 63.9 | 61.6 | 70.3 | 65.6 | 63.5 | 53.9 | 70.6 | 69.6 |
| UCB | 4 | 3 | 72.8 | 70 | 61.5 | 68.8 | 51.8 | 53 | 74 | 71.4 |
| UM | 21 | 21 | | 78.1 | | 60.6 | 23.8 | 31.4 | 48.8 | 44.2 |

*Note:* Shanghai ranking is a listing of world universities in order of total score. The highest ranking university gets 100 points and each university receives its score in relation to the #1 institution.

"Cited articles" is a measure of the number of publications by a university's faculty and staff that are listed in the Science Citation index-expanded and the Social Sciences Citation Index.

"Cited researchers" is the number of faculty and staff at a university that are listed by Thomson Reuters (publishers of SCI, SSCI, A&HI) as highly/frequently cited scholars.

"Per-capita academic performance" is a measure based on cited articles and cited researchers plus winners of Nobel Prizes and Fields Medals and number of articles in Nature and Science.

*Source:* SJTUIHE (2008).

ranked. In the comparable 2008 Shanghai ranking, Sichuan and Tianjin both made it to the bottom quintile of the top 500 list, while Peking and Tsinghua remained in the same mid-level grouping of institutions.

The most important aspect of table 12.7 is the absence of highly cited researchers at any of the five Chinese universities. While these institutions are increasing their scholarly output, the results are not yet considered to be of great significance to other researchers. In comparison, MIT has 82 highly cited researchers, Berkeley has 87, and Michigan has 68.

Of course, academic reputation is only one indication of an excellent university. Since the Shanghai ranking is frequently used as an indicator of university quality, however, it has developed a reputation perhaps greater than its value. Nonetheless, when political and academic leaders talk about the international competitiveness of their universities, the ARWU is usually the first ranking system they cite.

## Conclusion

The data presented in this chapter demonstrate that China's top universities are making great strides in terms of academic quality. The national government is investing heavily in China's top institutions through Project 985; universities that are not part of this project are stratified as national, regional, or local universities and not earmarked for special funding. The case study universities have used Project 985 monies to increase their research expenditures, publication rates, and status in international rankings.

The results vary, however, from campus to campus. While all of the universities in this study have made enormous progress, there is a real gap between the three case study institutions on the one hand, and PKU and THU on the other. There is a further gap between PKU/THU and the U.S. comparison schools, although on some measures THU is approaching the same level as U.S. universities.

One of the more telling statistics is the listing of "high citation" scholars. None of these Chinese universities have professors whose published research is highly cited by other researchers in the field. On the one hand, this finding is not surprising, given the recent entrance of Chinese higher education into the global competition for publication. On the other hand, it suggests that the current pressure for numbers of publications in international journals in some Chinese universities is a misguided priority. Perhaps fewer, but higher quality, articles would be a more suitable goal.

Our findings also suggest that world-class status may not be a realistic target for all members of the 985 Project. It might make more sense for some universities to make strategic identification of a small number of fields in which they can excel—a "peaks of excellence" approach—rather than striving for university-wide global recognition. This may already be happening, either unintentionally or intentionally, but is not apparent with institutional-level statistics.

Our data also show a priority placed on natural sciences and engineering, deemed vital for the nation's economic development, leading to an academic imbalance in much of Chinese higher education. The humanities and social sciences are not as productive in research terms compared with more technologically oriented disciplines. In addition, because the Shanghai index rankings emphasize the sciences, non-science disciplines contribute less to international stature and thus appear to be less valued on many campuses. The focus on science and technology is shortchanging other disciplines that are essential for creating comprehensive universities and a well-rounded society.

Similarly, the research emphasis has moved many Chinese professors away from teaching undergraduate students. The prestige granted to research also shortchanges the thousands of higher education institutions in China that are not striving for world-class status but are, in fact, educating the vast majority of students who will build a stronger nation. Chinese institutions are beginning to give greater attention to their students, however, in part because the Ministry of Education has an assessment tool that focuses specifically on undergraduate education.

Most Chinese university leaders know that they have not yet achieved world-class status by the measures cited in this chapter. In private conversations, they also say that they have not yet achieved qualitative measures of real international stature. As Philip Altbach (2004) points out, a world-class university has a number of intangible features, including internal self-governance and academic freedom. Chinese university officials recognize these shortcomings and seek greater autonomy.

Finally, it should be noted that the process of data collection for this chapter was challenging because so little information is publicly available. The challenge of gathering these statistics provides evidence of the lack of transparency in Chinese higher education. If Chinese universities wish to be full-fledged players in the world knowledge production system, the Ministry of Education, as well as individual institutions, need to build an open data system. If universities are to garner continued public support, Chinese institutions and the government agencies that control them will need to release meaningful data to the general public regularly.

# References

Altbach, Philip G. 2004. "Costs and Benefits of World-Class Universities." *Academe* 90 (1): 20–23.

Beijing Normal University. 2009. *History of BNU: Chronological Review.* Beijing: Beijing Normal University. http://www.bnu.edu.cn.

China's Ministry of Education (CMOE). 2008. Common Data Set of Higher Education Institutions, 2003 and 2007 [data provided to the authors]. Beijing. CMOE.

ISI Web of Knowledge. 2009. *Science Citation Index, Social Sciences Citation Index, and Arts & Humanities Citation Index.* Philadelphia: ISI Web of Knowledge. http://isiwebofknowledge.com.

Massachusetts Institute of Technology. 2008. *Financial Report 2007.* Cambridge, MA: Massachusetts Institute of Technology. http://vpf.mit.edu.

National Center for Education Statistics. 2009. *Integrated Postsecondary Education Data Set (IPEDS).* Washington, DC: U.S. Department of Education Institute of Education Sciences. http://nces.ed.gov.

National Science Foundation. 2008. *Survey of Research and Development Expenditures at Universities and Colleges.* Arlington, VA: National Science Foundation. http://www.nsf.gov.

Shanghai Jiao Tong University Institute of Higher Education. 2008. *Academic Ranking of World Universities.* Shanghai: Shanghai Jiao Tong University Institute of Higher Education. http://ed.sjtu.edu.cn.

Tsinghua University. 2009. *Introduction of Tsinghua University.* Beijing: Tsinghua University. http://www.tsinghua.edu.cn.

University of California, Berkeley. 2008. *Financial Report 2007.* Berkeley: University of California, Berkeley. http://controller.berkeley.edu.

University of Michigan. 2008. *Financial Report 2007.* Ann Arbor: University of Michigan. http://www.finops.umich.edu.

World Bank. 2008. "Global Purchasing Power Parities and Real Expenditures: 2005 International Comparison Program." Washington, DC: World Bank.

# Chapter 13

# Global Competition and Higher Education in Tanzania

*Ashley Shuyler and Frances Vavrus*

## Introduction

Higher education in the United Republic of Tanzania has changed dramatically since independence from Great Britain in 1961. The country's first tertiary education institution was established the same year with 13 students and became part of the University of East Africa in 1963 (Omari 1991; UDSM 2007). In 1970, the University of Dar es Salaam (UDSM) became the first national university, and it remained the only one until 1984 (Omari 1991). Today, more than 48,000 students study in Tanzania's eight public universities and 17 private higher education institutions (TMHEST 2005a).

Although Tanzania's higher education sector has expanded considerably since independence, it is still far below regional and global norms. For instance, its tertiary gross enrollment rate is only 1 percent compared to the continent average of approximately 5 percent and a global average above 15 percent (World Bank 2009). This low enrollment rate is largely a consequence of Tanzania's post-independence governmental policy that promoted universal primary and adult education (Vavrus 2003). As a result, secondary and tertiary education were highly restricted until the

1980s, and the effects can still be seen in that Tanzania has one of the lowest secondary school gross enrollment ratios—9 percent—in Africa (World Bank 2004). With fewer than 50,000 students enrolled in HEIs in a country of some 36 million people, the need for significant expansion in the secondary and tertiary sectors is evident, even while the best means of achieving this goal remain unclear.

Increasing access to post-primary schooling is only one of the challenges facing Tanzanian policymakers today. Another is the ongoing need to bolster the country's flagging economy: with per capita gross domestic product at US$1,300 (expressed in terms of purchasing power parity), Tanzania is one of the world's poorest countries (CIA 2008). This situation has influenced educational policy due to concerns that the low quality (and quantity) of secondary and tertiary education is hampering the country's ability to compete in the global economy. As a result, the Tanzanian government, with assistance from the World Bank and other international development institutions, recently launched a series of reforms designed to increase economic development by both expanding access to post-primary education and changing pedagogical practices. These reforms, based principally on what we are calling the discourse of *education for global competition* (EGC), are intended to make Tanzania more competitive internationally by replacing "traditional pedagogy" with "a culture favoring innovation" through the use of more applied, inquiry-based teaching and learning (World Bank 2009, 108). This change in instructional focus has been described as a shift from the fact-based, formalistic teaching common in Tanzanian schools to a social constructivist model that encourages problem solving, critical thinking, and student-generated knowledge (Stambach 1994; Vavrus 2009). Underlying these changes is the assumption that Tanzania will be able to improve educational performance among its students and, as a result, participate more fully in the global economy.

Educational reforms such as those currently promoted by the Tanzanian government and the World Bank may eventually improve economic productivity in Africa and make the continent's graduates more globally competitive. However, we argue that there has been insufficient attention paid to the impediments facing Tanzanian college graduates as they enter the global marketplace due to the structure of schooling at the primary and secondary levels. We suggest that the current EGC discourse minimizes the cultural, economic, and political dimensions of secondary schooling that impinge on the ability of higher education students to become engines of innovation. In this chapter, we examine three aspects of the secondary education system in Tanzania that impact the ability of college graduates to compete in a global environment: (a) the medium of instruction, (b) pedagogical practice, and (c) the national examination system. Although we discuss these issues discretely, they should be viewed as three interlocking strands of the same thread that weaves together primary, secondary, and tertiary education; one cannot be easily removed without altering the country's entire educational fabric.

To explore these three issues, we draw upon two separate but interrelated research projects. Ashley Shuyler carried out the first study in 2007 at four secondary schools in northern Tanzania as part of a study of high-stakes testing. Through interviews, questionnaires (N=499), and focus groups, Shuyler sought to understand the perceptions of Tanzanian secondary school students toward the high-stakes national examinations they take every two years. She also held 22 focus group discussions with teachers, parents, and recent secondary school graduates to discuss national examinations, tertiary education, and the economic competitiveness of college graduates. In addition, Shuyler conducted three interviews with educational policymakers in Tanzania's Ministry of Education and Vocational Training (TMEVT), National Examinations Council (NECTA), and Tanzania's Institute of Education (TTIE) to incorporate their perspectives.

Frances Vavrus completed the second study in 2006 and 2007, during which time she was engaged as a visiting professor and as an observer of daily life at a teachers college in northern Tanzania. This college was founded on the principles of active, participatory teaching and learning, making it an ideal site to explore the potential and the challenges of instituting pedagogical change. As part of this research, Vavrus lived and taught at the college and conducted in-depth interviews with six of her nine English methods students following their graduation. Several of the students had already begun teaching in secondary schools and could compare the methods they were taught at the college with those they found feasible in practice. Taken together, these two studies raise questions about how the goals outlined by the discourse of EGC in higher education can be achieved without fundamental reform of secondary schooling and teacher education.

Our attention to the education of secondary school teachers derives from two sources. First, teacher training constitutes a significant portion of the higher education sector in Tanzania; indeed, nearly 24 percent of annual bachelor's degree candidates at UDSM are studying to become teachers (TMHEST 2005b). It is these teachers who are preparing the next generation of secondary school students to enter Tanzania's institutions of higher education, so their perceptions and practices vis-à-vis policy reform are critical to understand. Second, pedagogical change has captured the attention of international financial institutions and policymakers seeking to improve Tanzania's competitiveness in the global economy.

Yet, the changes being advocated often reflect cultural and material conditions common in North Atlantic countries rather than in Africa. Our framework for analyzing these issues builds on Vavrus' work surrounding the *cultural politics of pedagogy*. This view suggests that educational policy reforms in countries like Tanzania that rely heavily on international development aid are acutely sensitive to economic and political forces within and beyond the nation-state and emphasize the cultural aspects of teaching that are not quickly altered through changes in policy. From the cultural politics of pedagogy perspective, one would question the implicit assumption that EGC should take the same form everywhere, regardless of context; instead, this view recognizes that the conditions of teaching vary considerably around the world and profoundly affect teachers' preparation for, and receptivity toward, constructivist pedagogy.

# The Discourse of Education for Global Competition

International development organizations, such as the World Bank, are increasingly embracing the view that significant reform in tertiary education is necessary for African countries to become more economically productive. From this perspective, institutions of higher education are seen as vital in the transformation toward competitive, knowledge-based economies (World Bank 2009). To achieve such reform at the tertiary level, the international community has encouraged the infusion of social constructivist pedagogical practices into secondary-level education, so that graduates will be prepared to participate in college classrooms that demand the use of high-level critical thinking skills. For instance, a 2009 World Bank report encourages African nations to seek economic growth by moving toward new techniques that promote growth and innovation. Such techniques, this report states, include "group work instead of lectures, problem solving rather than memorization of facts, [and] practical learning (field trips, attachments, internships) as a complement to theory" (108–109). With an increasing number of international donor organizations calling for this new pedagogical approach as a means of improving the economic competitiveness of African nations, ministries of education across the continent have worked to infuse national curricula with constructivist teaching methods (Lewin and Stuart 2003; Mulkeen et al. 2007; Chisholm and Leyendecker 2008). In particular, the international community has encouraged these nations to use teacher training as the point of entry for new pedagogical methods. International organizations see reforms at the secondary level as intimately associated with the degree to which college-level students will later be prepared to participate in globally competitive institutions of higher education.

In response to these calls by the international development community to transform secondary-level education and modify teacher training curricula, the Tanzanian Ministry of Education initiated the Secondary Education Development Plan (SEDP) in 2004 through a US$123.6 million loan and a US$26.4 million grant from the World Bank. In its first two years, SEDP succeeded in doubling the gross and the net enrollment rates in Tanzania's secondary schools and dramatically increasing the number of classrooms and the size of the teaching force in the country (Makwasa 2006; Woods 2007). At the same time, the Tanzanian government has moved to satisfy its international constituencies by including provisions in SEDP to improve teacher education so as to "discourage the excessive use of memorization in the classroom" in order to make tertiary-level training programs "less academic, more balanced and more practical" (Mkapa 2004, 9–10). These measures, designed to ensure that Tanzania's secondary- and tertiary-level institutions are competitive with their global counterparts, were formalized in 2005, when TTIE initiated a parallel program to SEDP to revise its curricula from a content-based approach that focuses on the teacher to one directed at students and their emerging competencies (TMEC 2005).

The EGC and constructivist discourses were evident in our interviews with Tanzanian policymakers, who share many of the same views as their counterparts in international development institutions. However, their views on education reform were inflected by references to the country's cultural and economic conditions and to a global climate they viewed as "hostile" for Tanzanians. They situated the concept of the individual entrepreneur within a social context that reflects the country's socialist history in which a collective sense of "self-reliance"[1] was once cultivated. One TTIE official described this international climate and the need for curricular change:

> The environment is very hostile now. Tanzania needs people who are creative, and who are problem-solvers. If people aren't creative, they won't survive. Now things are changing, the country needs people to perform. There is competition from Uganda, Kenya, and the USA. Globalization is coming. We want to create thinkers. We don't want rote learners now, but instead people who can change our environment. The private sector is coming up, and it wants people who can make changes, not people who are waiting for procedures. (Interview, June 28, 2007)

Another TTIE official discussed the importance of changing Tanzanian curricula to reflect EGC discourses, but he related it specifically to the

Tanzanian concept of self-reliance, in which the individual's "entrepreneurship" should also contribute to the ability of the community to employ its members:

> Tanzania has made this switch because it wants its graduates to perform. [These reforms] are connected with the idea of entrepreneurship: that graduates should be able to be self-reliant and dependable, not just able to remember the facts that he or she has learned, but can be self-employed and can employ others. Today if you graduate from university, you must be able to create a job for yourself, and for others. (Interview, June 28, 2007)

Thus, educational policymakers in Tanzania espoused the discourse of EGC circulating within the international development community in their belief that Tanzania's modified curriculum will create citizens who are better equipped to think creatively and act entrepreneurially in the face of the expanding, global free market. Yet, they also indicated that entrepreneurship has a somewhat different meaning in Tanzania, where the limited availability of capital for most individuals means that the "entrepreneurial spirit" most often takes the form of self-reliance—meaning, in this case, relying on one's connections in the community to enable oneself to generate an income. The discourse of EGC does not take into account such global variations in the competitive landscape in which college graduates find themselves. In the section that follows, we look more closely at three features of Tanzania's secondary education system—language, pedagogy, and examinations—which, ultimately, limit college graduates' ability to engage in the single global economic landscape envisioned by the international development community.

## Multiple Languages for a Single Competitive Landscape

Tanzanian policymakers and students alike recognize that the language of global competition is English. Tanzanian teachers and student teachers, however, expressed concern that the use of English as the medium of instruction in secondary schools inhibited their ability to use constructivist methods and limited the ability of students to learn effectively.

Speaking to the importance of learning English at the secondary level, one policymaker at the Ministry of Education argued:

> The world is now a small village because of the free market, so after secondary school, students must have increased language competencies.... If you

go to the airport, you need English. You need English to communicate with the outside world.... In the globalized world, people are moving from place to place seeking jobs. We need English so that people can interact easily. (Interview, June 26, 2007)

Likewise, when students were asked their opinion in focus group discussions about the most important subject learned in secondary school, the most common response was "English," and it was almost always linked discursively to global competition, with explanations such as the following offered:

With regard to English, we have entered the era of globalization, and all of the transactions are conducted in English. So, if we [Tanzanians] do not prioritize English as other subjects, our students will be ill-equipped to compete in the global market and may be considered misfits. (Focus Group, July 25, 2007)

Both policymakers and students overwhelmingly affirmed that English must be learned if Tanzania wishes to become globally competitive.

In contrast, however, teachers and student teachers noted that reliance upon rote memorization in Tanzanian classrooms is exacerbated by the use of English as the medium for instruction in secondary schools. One teacher commented that it was nearly impossible to use constructivist methods in the classroom because "most students memorize materials which they do not understand since English is not their mother tongue" (Focus Group, July 23, 2007). Because students are required to use a language in which they are not fluent, teachers argued that they resort to memorization in place of methods that might promote greater understanding and, as a result, their students are not prepared to participate later in higher education or to contribute to economic life in a meaningful way. As one teacher complained, "the problem with Tanzanian education is that most of the students memorize for exams but practically in life they have nothing" (Focus Group, August 2, 2007). Similarly, a student teacher elaborated on this problem:

I am glad I was born in town, but a Pogoro [ethnic group in rural Tanzania], maybe, when he or she is born, he will first learn Pogoro and then Kiswahili at school, primary school, and at secondary school she or he will have to speak English.... Now, it's a problem—a very, very big problem. Because now you are having to transform your knowledge of Pogoro into Kiswahili and then into English. So it's very difficult. So you will find pupils cramming how to write "crust" while they don't know how it's written or sometimes they don't know even the meaning. (Interview, February 1, 2007)

These teachers' and student teachers' views suggest that while the use of English in Tanzanian secondary schools may be intended to enhance participation in the global marketplace, it also induces students to memorize in place of comprehending and analyzing critical information, therefore hindering the efforts of teachers to use constructivist pedagogy. Even though the discourse of education for global competition makes English a logical choice for the classroom, it fails to acknowledge the complexity of language proficiency in a nation like Tanzania.

# Pedagogy and Poverty

While today's Tanzanian teachers are themselves largely products of the same rote learning system that is still prevalent in Tanzanian classrooms, many have demonstrated in teacher training colleges their ability to master the theoretical principles of the constructivist teaching techniques that form the core of discourses of EGC. However, most graduates of Tanzanian teacher training institutions struggle to implement the methods required by the new curricula once they are in the classroom. Observations of secondary-level classrooms and interviews with teachers and student teachers make it clear that the policies recommended by the international development community and subsequently enacted by the Tanzanian government have not translated as expected to the secondary school setting. According to teachers, reforms toward constructivist teaching have not succeeded largely because the material environment in Tanzanian schools has not been given adequate consideration. Indeed, the majority of teachers trained in Tanzanian colleges continue to use teacher-centered practices in the classroom both because they have not had an opportunity to practice student-centered methods sufficiently and because material challenges make teacher-centered, fact-based instruction the most logical pedagogical choice in large, underresourced classes.

Teachers and student teachers consistently remarked that they felt constrained in their ability to use constructivist techniques when placed in crowded classrooms with very few resources. For example, in a focus group discussion with a group of secondary school teachers, one teacher described his desire to utilize a "participatory method" in which he could "involve [students] in searching for the materials on a certain topic," but he emphasized that there was no library or other reference materials that students could use (August 2, 2007). A student teacher made a similar comment about the methods he was taught at his teacher training college:

> [W]e were taught different teaching methods, games, to apply in [the] teaching process. We made some displays, posters, and I think that I will go to apply them in the schools. But something I worry [about] is that some of the schools haven't any teaching and learning materials. Yeah. So it will be difficult to implement other methods. (Interview, February 1, 2007)

With teaching aids limited, in many cases, to a chalkboard and a single textbook, these teachers find it nearly impossible to implement the constructivist methodologies advocated by policymakers in which students plan and carry out their own inquiries.

Likewise, teachers are often challenged to utilize participatory techniques due to classroom space limitations. While the average class-size in secondary school is supposed to be maintained at 30 pupils per teacher (SEAI 2007), it is not uncommon to find classrooms packed with 50 or more students, with several students sharing a single desk. In classrooms such as these, group work, experiments, and simulations are much more difficult. For example, one student teacher reflected on his experience in a large class in an urban school in which he wanted to play a game he had been taught at college using flashcards. He explained that some participatory techniques like this one could not be used: "I remember once I taught in [name of school]. And you find the class with 110 students. Really the class, it, maybe you can't apply those things—certain techniques" (Interview, January 30, 2007). This example highlights the challenges that limited materials and large classes pose for secondary school teachers. Tanzanian teachers generally have to work harder to implement active, student-centered methods than teachers in resource-rich schools, making it more difficult for their pupils to attain a competitive edge.

# High-Stakes Exams as a Barrier to True Competition

In addition to the significant material challenges Tanzanian teachers face in their classrooms when trying to implement constructivist pedagogy, they must also negotiate the tension created by the system of national examinations that demands the memorization of vast quantities of information. Tanzania's secondary-level examination system is composed of three high-stakes tests, taken after the second, fourth, and sixth years of secondary school. Characterizing these exams as high stakes is not an exaggeration in that the historical pass rate of the fourth-year secondary school exam effectively prevents 75 percent of students from continuing to their final two years of secondary education (MEVT 2006). A great deal of memorization is required for these tests because students are examined in multiple choice and sentence-completion formats in seven to ten subjects, each of which covers four years' worth of information.

In focus group discussions, secondary school teachers spoke extensively about the ways in which the content, format, and high-stakes nature of these tests influence their daily classroom lessons. For example, one teacher noted that the "national exam is all that matters" because students "know that the exams will classify them to their different abilities: who is gold, who is iron, who is silver." As a result, he continued, "Many students learn how to pass exams, and not simply to gain knowledge itself. Even teachers just learn how to help students pass the exam" (Focus Group, August 6, 2007). The high-stakes nature of the fourth-year exam induces teachers to emphasize only those topics tested, often at the expense of other, perhaps equally important, subjects.

In addition to course content, the examinations shape the manner in which teachers conduct lessons. Because the exams demand an extraordinary amount of memorization, many teachers believe that it would be negligent to devote large amounts of time to allowing students to discover answers themselves when they could instead be provided with the information that they will need to pass. One teacher explained his decision not to use inquiry-based methods of teaching in the following way: "I will have to teach them because education in Tanzania is examination-based in such a way that if a student does not pass [he] is counted as if he did not study or did not understand and so missing the possibility to continue with Form Five" (Focus Group, July 23, 2007). Teachers often felt that they did not have the luxury or freedom to utilize the student-centered methods that are supposed to help produce globally competitive youth. These teachers believe, instead, that to do so would leave their students ill-equipped to compete on their national exams and therefore put them at a disadvantage when attempting to gain access to the limited, highly coveted spots in higher education.

In this way, the Tanzanian examination system cultivates a tendency on the part of both teachers and students to focus on mastering the large amount of content tested on the examination instead of on those skills promoted by the constructivist discourses of EGC. Although teachers and students alike recognize such memorization has only limited application later in life, they continue to rely upon these techniques because of the all-important nature of national examinations in determining educational and professional futures.

## Conclusions

The Tanzanian teachers, students, and policymakers who contributed to our studies shared the broad goal of encouraging and enhancing higher education as a means of furthering economic development at the national and/or individual levels. However, their views also indicated that there are different degrees of recognition of the challenges associated with changing the content and practice of secondary and tertiary education as a means of improving graduates' ability to compete in a global economy. Policymakers generally felt that using English and constructivist methods at the secondary level are both essential and realistic measures to achieve the goals outlined by the EGC discourse. While teachers and student teachers did not disagree with these goals, they expressed great difficulty in actually implementing the student-centered methods recommended by international institutions when faced with limited resources, an examination system that does not reward such efforts, and students who are not fluent in the language of instruction. Thus, when placed into the "real world" of Tanzanian secondary school classrooms, constructivist pedagogy is usually ignored or, at best, transformed into formalistic practice with only hints of student-centered teaching and learning. These findings reinforce the applicability of the framework offered by Vavrus' *cultural politics of pedagogy*, which recognizes the cultural, economic, and political forces that limit the transformation of instruction on constructivist pedagogy at teachers colleges into practice in Tanzanian classrooms. This understanding calls into question the international effort to urge—or even require— African governments to promote constructivist pedagogical approaches when other structural conditions of schooling may not have first been appropriately addressed or altered.

Even though Tanzania has worked to reshape its secondary and tertiary education policies to reflect the discourse of EGC, our research indicates that it will be quite some time before its students are adequately prepared for participation in globally competitive institutions of higher learning. In this chapter, we have shown that secondary school students' limited skills in English, teachers' limited classroom resources, and a national examination system that demands memorization over critical thinking affect the abilities and dispositions of students and teachers for the radically different kind of classroom environment desired by government and international development organizations. Only when these material, political, and cultural constraints are accounted for in the effort to create more internationally competitive HEIs will these reforms lead to the development of more internationally competitive students in Tanzania.

# Note

1. Here, "self-reliance" (or *kujitegemea* in Swahili) refers to the set of philosophies and policies established by Tanzania's first president, Julius Nyerere, in his 1967 *Education for Self-Reliance* address. In this context, self-reliance refers to the self-sufficiency of an individual or community through cooperative efforts. Nyerere emphasized that this concept implied "the obligation to serve, as well as be served, and to cooperate, rather than compete" (Nyerere in Lema et al. 2006, 189).

# References

Chisholm, Linda, and Roman Leyendecker. 2008. "Curriculum Reform in Post-1990s Sub-Saharan Africa." *International Journal of Educational Development* 28 (2): 195–205.

Central Intelligence Agency. 2008. *Rank Order—GDP—Per Capita (PPP)*. Washington, DC: Central Intelligence Agency. https://www.cia.gov.

Lema, Elieshi, Issa Omari, and Rakesh Rajani. 2006. *Nyerere on Education: Selected Essays and Speeches, 1961–1997*. Dar es Salaam: HakiElimu.

Lewin, Keith M, and Janet S. Stuart. 2003. "Researching Teacher Education: New Perspectives on Practice, Performance and Policy." In *Multi-site Teacher Education Research Project (MUSTER): Synthesis Report*. London: Department for International Development.

Makwasa, Makwasa. 2006. "267,224 to Miss Out in Govt. Secondary Schools." *The Citizen*, December 20: 1–2.

Mkapa, Benjamin. 2004. *Mpango wa Maendeleo ya Elimu ya Sekondari [Secondary Education Development Programme]*. Dar es Salaam: HakiElimu.

Mulkeen, Aidan, David Chapman, Joan G. DeJaeghere, and Elizabeth Leu. 2007. *Recruiting, Retaining, and Retraining Secondary School Teachers and Principals in Sub-Saharan Africa*. Washington, DC: The World Bank.

Omari, Issa M, 1991. "Innovation and Change in Higher Education in Developing Countries: Experiences from Tanzania." *Comparative Education* 27 (2): 181–205.

Secondary Education in Africa Initiative. 2007. *Synthesis Report: Executive Summary*. Washington, DC: World Bank. http://siteresources.worldbank.org

Stambach, Amy. 1994. "'Here in Africa, We Teach; Students Listen': Lessons about Culture from Tanzania." *Journal of Curriculum and Supervision* 9 (4): 368–385.

Tanzania's Ministry of Education and Culture. 2005. *Information and Computer Studies Syllabus for Secondary Schools, Form I-IV.* Dar es Salaam: Ministry of Education and Culture.

Tanzania's Ministry of Education and Vocational Training. 2006. *Basic Education Statistics in Tanzania (BEST): 2002ñ2006 National Data.* Dar es Salaam: Ministry of Education and Vocational Training.

Tanzania's Ministry of Higher Education, Science, and Technology. 2005a. *Grand Total of Students Enrolled in HLIs for Year 2000/2001–2004/2005.* Dar es Salaam: Ministry of Higher Education, Science, and Technology. http://www. msthe.go.tz (under construction).

———. 2005b. *Student Enrollment General.* Dar es Salaam: Ministry of Higher Education, Science, and Technology. http://www.msthe.go.tz.

University of Dar es Salaam. 2007. "Institutional Self-assessment of the University's Mission in Relation to the Civic Role in and Social Responsibility to Society, With a Special Focus on Two Self-initiated Global Projects." Unpublished report, University of Dar es Salaam, Dar es Salaam.

Vavrus, Frances. 2003. *Desire and Decline: Schooling Amid Crisis in Tanzania.* New York: Peter Lang.

———. 2009. "The Cultural Politics of Constructivist Pedagogies: Teacher Education Reform in the United Republic of Tanzania." *International Journal of Educational Development* 29 (3): 303–311.

Woods, Eric. 2007. "Tanzania Country Case Study." Country Profile Commissioned for the *EFA Global Monitoring Report 2008: Education for All by 2015: Will We Make It?* Paris: United Nations Educational, Scientific and Cultural Organisation.

World Bank. 2004. *Program Document for a Proposed Adjustment Credit in the Amount of SDR 82.7 ($123.6 million) and a Grant in the Amount of SDR 17.7 ($26.4 million) to the United Republic of Tanzania for a Secondary Education Development Program.* Dar es Salaam: World Bank, Africa Regional Office.

———. 2009. *Accelerating Catch-up: Tertiary Education for Growth in Sub-Saharan Africa.* Washington, DC: World Bank. http://siteresources.worldbank.org.

# Chapter 14

# Soft Power Strategies: Competition and Cooperation in a Globalized System of Higher Education

*Joseph Stetar, Colleen Coppla, Li Guo, Naila Nabiyeva, and Baktybek Ismailov*

The rise of the knowledge economy (Drucker 1993) and dramatic shifts to the global competition landscape are changing the nature of higher education, with universities in many of the most developed nations trading their traditional roles of addressing regional or national needs for increasing political, cultural, and economic influence around the world. It is no longer sufficient to be a leading national university; leading universities now see the need to demonstrate a substantial global reach as well.

Manifestations of these competitive drives among universities abound. Efforts at brand extension are evident, for instance, in the establishment of branch campuses in Qatar by New York University, Cornell, Georgetown, and Carnegie Mellon; the State University of New York at Buffalo's outpost in Singapore; and the many U.S. universities which are competing with their British, Australian, and Canadian counterparts to establish programs in China. The intensity of global higher education competition is particularly evident when one turns to the back pages of *The Economist* magazine, where universities take out glossy advertisements proclaiming their global reach.

However, there is another, perhaps less recognized, dimension to global competition among universities, evident in the increasingly central role universities have assumed in what Joseph Nye (2003, 2005) describes as the projection of "soft power." Soft power—which Nye conceived as a political strategy used to foster appreciation and acceptance of a nation's culture and values—draws on the subtle effects of culture, values and ideas, in contrast to the more direct, tangible measures that hard power encompasses (such as military force or economic influence). Soft power seeks to gain appreciation for a nation's culture organically, as contrasted with hard power's more forceful nature. It uses a combination of attraction and persuasion to reach its expected outcome; hard power threatens, while soft power encourages. This attraction and encouragement can stem from a nation's ideals, traditions, religion, art, language, or a combination of these entities. Soft power flourishes when the nation projecting that power is well-respected on an international scale. Nye suggests that a nation with effective soft power initiatives has little need for immense, excessive, costly, and often insufficient hard power strategies (Nye 2005).

In a world intimately connected by technology and fueled by a growing knowledge economy, nations and various religious, cultural and linguistic groups increasingly see universities as powerful instruments for projecting soft power and expanding their spheres of influence. There is, for example, considerable competition between Russian-language private universities in Eastern Ukraine and Ukrainian-medium private universities in the western portions of that nation. There is also competition among Western-sponsored Christian universities, and those sponsored by the Orthodox Church (Stetar et al. 2007). In Middle Asia, nations such as the Kyrgyz Republic are witnessing an emerging competition between liberal Western-style universities, traditional Russian higher education, and more fundamentalist Muslim institutions. Furthermore, China is increasingly using its higher education institutions to project soft power in a number of strategic regions.

In the remainder of this chapter we examine how the United States and China are using universities to project soft power and gain global advantage. Additionally, we illustrate how religious, cultural, and linguistic groups are utilizing universities to project soft power in two former Soviet Republics: Azerbaijan and Kyrgyzstan. Collectively, these transitional countries are unique examples where particular assets, strategic foreign policy interests, and a positioning for soft power projection converge through the vehicles of language, religion, and higher education.

# United States: Historical Projector of Soft Power

A nation's soft power capacity rests in its culture and values when these are attractive to others, and when they are realized both domestically and internationally. For soft power to exist, a nation's values must be considered moral and legitimate. These soft power characteristics can be conveyed through a nation's higher education system, both within a country's own borders and by delivering a nation's higher education abroad. This delivery happens physically, through overseas campuses and international institutional collaborations, but also theoretically, such as when international students are educated in one country and return to their homeland or another nation, bringing with them the perspective of the country where they were educated. Their educational experiences will inevitably shape how they see the world, and in doing so, further influence the nation that educated them.

University-level U.S. educational opportunities are provided to citizens of many nations both abroad and within the borders of the country. Through Fulbright and related programs or bi-lateral cooperative agreements between universities, the United States has used higher education as a vital element in its foreign policy strategy, and has created a legacy which promotes its ideals, values, and culture both domestically and internationally.

An example of this can be drawn from the Cold War, which was waged with a blend of hard and soft power. Academic and cultural exchanges between the United States and the former Soviet Union, starting in the 1950s, played a vital role in advancing U.S. soft power (Nye 2003) and calming tensions between these nations. These same cultural experiences may have also contributed to the ultimate downfall of the Soviet Union. While pessimists argued that Russians would use their time studying in the United States to uncover scientific secrets, educational exchanges positively impacted the world view of Soviet students who studied in the United States. It ultimately helped them to appreciate and accept some U.S. political ideals and therefore influenced their beliefs and actions when they returned home.

Historically, the use of U.S. higher education as a vehicle for soft power projection has been part of an overall strategy in U.S. foreign policy to create allies and understanding for U.S. ideals across the globe. It has benefited from a wide-reaching, open culture with attractive values, such as freedom, tolerance, and open-mindedness. However, after the terrorist attacks of September 11, 2001, U.S. citizens were justifiably concerned about homeland security and the need for safety measures vis-à-vis foreign students. Cautious measures quickly became severe with the implementation of draconian visa and screening procedures. While the situation has improved dramatically in recent years, the United States continually needs to be reminded that its influence in the world has long been tied to an educational system that is both desirable and available to foreign nationals. Closing the doors to educational opportunities and exchange, equals closing the doors to a primary source of soft power for the United States.

Soft power does not simply exist by a nation's choice to deliver it. When a nation strategically uses vehicles such as higher education to project soft power, a nation's soft power can grow or dissipate based upon the examples it sets; it can either magnetize or repulse (Nye 2005). If a nation's political stances are well known and respected, the nation must project those political stances and live by its own example. Perception is an important aspect in the capacity of soft power. If opinions about a country deteriorate, this can greatly impact that country's ability to successfully project its soft power through any means. The reputation of a country, both domestically and internationally, is crucial, and part of this perception includes, but is not limited to, the desire of others to pursue a degree in a nation's system of higher education. A decline in a nation's soft power can also impact the status of its education system and ultimately make a nation less competitive in the global academic market and economy. Domestic or foreign policies that appear to be insincere, contradictory, or self-serving lessen a nation's soft power and, therefore, that nation's overall global influence.

The United States has certainly experienced a lessening of its soft power since September 11, 2001. There has been a steep decline in perceptions of the attractiveness of the United States, both in terms of its receptiveness to foreign students and its unpopular foreign policies, as measured by polls taken in a 2003 worldwide survey (Pew Research Center 2003). Respondents with unfavorable views of the United States most often said they were reacting to President George W. Bush's administration and its policies rather than the country's people and overall culture, suggesting that while it may not be as powerful as before, the United States has been able to retain soft power in certain areas. Respondents in most nations noted they still admire the United States for its technology and the entertainment industry, but large majorities in many countries said they did not favor the country's growing global political influence.

Collectively, these results raise the issue of whether the use of universities to project soft power is seen as benign or as an overly aggressive effort to impose one nation's values on another. Many observers agree that higher education in the U.S. projects significant soft power. Former U.S. Secretary of State Colin Powell said in 2001, "I can think of no more valuable asset to our country than the friendship of future world leaders who have been educated here" (Nye 2008, 14). A 2007 poll of 2,536 global leaders reported that 88.5 percent earned at least one degree in Western universities, and nearly half of these institutions were located in either the United States (27.4 percent) or the United Kingdom (18.8 percent) (Lee 2007). Although U.S. soft power may have faded in recent years, other observers have suggested that the country's soft power is still greater than its hard power, thus supporting the notion that its legacy of soft power is resilient and will endure, even in light of internationally unpopular U.S. political actions or stances (Joffe 2001).

While soft power projection continues to be important for the United States, other nations are also making significant investments in this area. Recent initiatives by China, for instance, suggest it is increasingly relying on soft power initiatives to expand its global influence.

## China: The New Competitor

Concurrent with China's rise as a global economic power has been an effort to extend its influence through the strategic use of soft power resources (Kurlantzick 2007). Since the late 1990s, a more coherent Chinese foreign policy can be found toward areas it sees as strategically important, promoting a higher level of internationalization and indicating a desire to compete for influence with other economic powers such as Japan and the United States (Yoshihara and Holmes 2008).

Recognizing the centrality of language in increasing its cultural attractiveness, Beijing has introduced a series of strategies to promote Chinese language and culture around the world (Ding and Saunders 2006). In 1987, the Chinese government established the National Office for Teaching Chinese as a Foreign Language (NOTCFL). Affiliated with China's Ministry of Education, NOTCFL was charged in 2004 with the mission of establishing a global network of overseas Chinese learning centers called the Confucius Institutes, which combine higher education with the appeal of Confucianism (Yang 2007). Since the establishment of the first institutes in 2004, they have quickly exhibited a global reach, with 262 operating in 75 countries: 93 in Europe, 72 in Asia, 68 in America, 21 in Africa and eight in Oceania. Current plans are to have 500 Confucius Institutes by the end of 2010 (NOTCFL 2008).

The Confucius Institutes have proven quite effective in promoting Chinese language and culture as well as strengthening the "understanding, opportunities and bonds among individuals, enterprises, communities and institutions" overseas with the People's Republic of China and the global Chinese diaspora (Ding and Saunders 2006, 16). Since the late 1990s, China has been the largest source of economic assistance to Southeast Asian countries, such as Laos, Vietnam, Burma, Cambodia, the Philippines, and Indonesia. Aid was initially focused on infrastructure and large prestige projects in the strategic areas, but now has been largely re-directed to soft power initiatives such as educational and cultural programs. Indeed, soft power initiatives have earned China greater respect in the region, as governments feel less threatened in their relations with the country, thus allowing China to play a more active role in regional organizations such as the Association of Southeast Asian Nations (ASEAN) (Lum et al. 2008).

The establishment of Confucius Institutes appears to be closely tied to China's strategic interests. For example, the institutes teach simplified Chinese characters, which are Beijing's preferred version, rather than the traditional characters that are used in Taiwan. This strategy is perceived as helpful in advancing Beijing's goal of marginalizing Taiwan in the battle for global influence (Ding and Saunders 2006). In addition, China's expectation of growing future dependence on energy imports has brought the Confucius Institutes to places such as Kazakhstan, Russia, Venezuela, Nigeria, Iran, Saudi Arabia, and Canada. Consistent with the use of the institutes and their linkages with indigenous universities to further Chinese strategic interests, there are, for example, nine institutes in Russia, five in Canada, two in Nigeria, two in Kazakhstan, one in the Sudan, and one in Iran (NOTCFL 2008). Although the NOTCFL did not disclose any energy policy interests in establishing the Confucius Institutes, it is believed that their appearance in these countries is not serendipitous, but rather a sign of China's determination to utilize the institutes and their linkages with indigenous universities to project soft power in areas of strategic importance.

The Confucius Institutes are often considered to be similar to the British Council, German Goethe Institutes, and France's *Alliance Française*, in that they are government organized and funded to promote the country's language, education and culture through various language and cultural programs, such as teaching Chinese, training for Chinese teachers, hosting Chinese speaking competitions, and providing consultancy services for students who wish to study in China.

However, the Confucius Institutes are distinctly different from British or German agencies in the way in which they engage in their host countries (Adams 2007). The European organizations tend to locate their offices in commercial areas and have not generally been as closely integrated into their host societies via institutional partnering as their Chinese counterparts. In contrast, the Confucius Institutes are incorporated into leading universities around the world, and simultaneously link the host university with key Chinese universities through supportive twinning arrangements (Yang 2007). For example, the Confucius Institutes established at Moscow State University were an outcome of a twinning arrangement between Moscow State University and China's Peking University. As the host university,

Moscow State houses the institute, while Peking University is responsible for providing teaching staff and materials. Similarly, the London School of Economics and Political Science operates a Confucius Institute in collaboration with China's Tsinghua University, while *Shanghai Jiaotong University* has partnered with Purdue University (Confucius Institute Online 2009). These collaborations provide the benefit of integrating the institutes into universities having a vested interest in supplying them with support (Yang 2007).

In supporting the Confucius Institutes, the Chinese government spends only about US$37 million annually. However, the institutes have received significant global publicity, testifying to their success despite their meager budget (Lai 2006). Chinese leaders have also given strong backing for the project, with both Chinese President Hu Jintao and Vice-President Xi Jinping visiting overseas Confucius Institutes (CCTV 2009).

China's use of higher education to further its national interest has not been limited to the Confucius Institutes. Its soft power projection is readily apparent in developing nations such as Africa, where China has dedicated itself to human resource development in the region. Annually, over 1,500 African students are awarded scholarships to study in China, and many Chinese universities have established formal partnerships with African institutions. Africa is also receiving technical aid, as China expands and strengthens its influence in the region. Medical, agricultural, and engineering teams, often drawn from universities, support national health programs in Africa, such as treatment of AIDS patients (Yang 2007).

Beyond soft power projection in developing nations, educating the elite of other nations is an important soft power strategy that has a long tradition in the United States. It now appears that China is determined to compete with the United States in this role (Whitney and Shambaugh 2008). China is recruiting students from around the world to study in Chinese universities, and the future generations in these nations, if Chinese efforts are successful, will be sensitized to Chinese viewpoints and have a greater appreciation of Chinese culture, language, history, politics and religion. Statistics show that increasing numbers of international students are studying in China, with the enrollment of foreign students from 178 countries studying for advanced degrees at China's universities more than tripling in 2004, to 111,800 from 36,000 in recent decades (CMOE 2008a). In total, China attracted more than 1.23 million students from abroad in the last 30 years (CMOE 2008b).

Although China does not have the long history that the United States has in using its universities to project soft power, it is evident that China has been putting great effort forth in this area and is moving forward with specific tactics to further increase its soft power in areas of strategic importance. If China's strategy continues to be carried out successfully, China will increasingly become an appealing and powerful entity on the world stage of business, commerce and culture, creating a new legacy of soft power that is its own.

## Kyrgyzstan and Azerbaijan: Battleground States

Kyrgyzstan and Azerbaijan are nations that are strategically suited for soft power projection from key global forces, and thus create a battleground of sorts for nations seeking influence there. From the late nineteenth century throughout most of the twentieth century, Central Asia and the Caucasus area remained under Russian and then Soviet control, isolated from the rest of the world. All of this radically changed with the breakup of the Soviet Union at the end of 1991, and the weakening of the Commonwealth of Independent States as an integrative structure.

These regions continue to be a likely battleground for outside powers. Russia is attempting to reclaim its status as the regional superpower, while China is using its new economic, political, and cultural power to build its image as a responsible world leader (Lawrence 2009). Turkey and Iran have also sought to expand their influence, based on historical and cultural ties. Meanwhile, certain Arabic countries, such as Saudi Arabia, Kuwait, and Qatar, desire to exert an influence through the revival of Islam in the region. Finally, a relative newcomer—the United States—is making its appearance, positing Central Asia and the Caucasus States as integral to its efforts to confront global terrorism (Burghart 2007). This focus has propelled the United States to expand its military and strategic footholds in the region.

Because soft power can directly affect all sectors of a society—namely education, culture, politics, ideology, religion, and industry—a number of key powers appear to be engaged in serious internal debates about the importance of using not only military strength and other hard power initiatives but soft power initiatives as well, to achieve longer-range goals in Kyrgyzstan and Azerbaijan. Higher education in particular is utilized by nations, religions, linguistic groups, and ethnic groups to gain influence. Below we discuss the ways in which language and religion interact with higher education to serve as a means of projecting soft power.

The emergence of English language universities suggests that Russian appears to be slowly losing its linguistic supremacy among the educated classes of Azerbaijan. During the Soviet period, educated families through-out the Russian Empire viewed the language of Pushkin and Tolstoy as essential to getting ahead and being considered fully "civilized." During the Soviet era in Azerbaijan, the Russian language prevailed over Azeri in everyday situations. People who had only limited Russian tried to use it. Almost all documents, meetings and conferences were held in Russian and if someone could not speak Russian at a Communist Party meeting they were rarely given the floor. Speaking Russian was critical for the upwardly mobile classes, and throughout the 1970s and 1980s it emerged as the first language for the children of the intelligentsia and professional middle class (Heyat 2002).

Since Azerbaijan gained independence, however, Azeri students have increasingly chosen English, not Russian, as their first foreign language, with French or German as their second foreign language of choice. Azeri students began to exhibit a strong desire to integrate into European or U.S. communities rather than cultivate their Russian language skills. The increasing need for English-speaking graduates to work with the growing number of Western enterprises established in Azerbaijan has also created a new demand for trained English-speaking graduates. Thus, for many members of the elite in Azerbaijan, the cultural center of gravity is no longer Moscow.

In Kyrgyzstan the picture is a bit different. As in Azerbaijan, the Russian language dominated during the Soviet era, and the first university programs in the Kyrgyz language appeared only in the early 1990s after independence. Currently, approximately 68.0 percent of students study in Russian, 28.0 percent in Kyrgyz, 2.0 percent in Uzbek, 1.1 percent in English and 0.3 percent in Turkish (National Statistical Committee 2008). Unlike in Azerbaijan, Russian remains the dominant language of higher education in Kyrgyzstan and is the students' portal to the outside world.

Since Russian has cemented itself as the dominant academic language in Kyrgyzstan, Moscow continues to provide post-Soviet schools and universities in the region with texts for a variety of courses in Russian, and thus continues to influence its "soft power" through language. However, it seems the number of university students learning in English will increase. The pull of English as a world language, and the attendant access to information the language affords university students, suggests that its growth in Kyrgyzstan is inevitable. There is also the expectation at the Kyrgyz Ministry of Education (KMOE) that Turkish language instruction will increase slowly in future. The KMOE is quite clear in its policies to further the development of the Kyrgyz, Russian, and English languages. Communication occurring in a common language between participants to pursue diverse interests is one strategy of soft power projection.

Perhaps nowhere in Middle Asia is the competition between universities supported or sponsored by foreign organizations or nations felt more clearly than in Kyrgyzstan. With strategic importance to both Russia and the United States, it is not surprising that these nations have a definable higher education footprint in Kyrgyzstan. The Kyrgyz-Russian Slavonic University and the Kyrgyz-Russian Academy of Education, as well as seven branch campuses of different Russian universities dispersed across the country, compete with the Western-style American University of Central Asia for students and influence. Also seeking to extend national and cultural influences in Kyrgyzstan are Manas University and Ala Too University (which have roots in Turkey), and Mahmud Kashgary University, which receives considerable support from Kuwait.

The foreign university influence in Azerbaijan is much more muted. Currently only two foreign universities have established institutions: Qafqaz University was founded in 1993 by Turkish businessmen, and more recently, an affiliate of Moscow State University have been established in Baku.

In addition to higher education, religion and ideology are also seen by some as potent vehicles for projection of soft power in these areas. In the Communist Bloc, religion was officially stigmatized as the "opiate of the people" (Reychler 1997). A growing impact of religious discourses on international politics is obvious today, and educational institutions are major sources of soft power used by religious institutions and governmental organizations to expand ideological influence.

Ideological disorientation, a search for national and cultural identity, and tensions with Armenia are three major factors that define the place of Islam in modern Azerbaijan. With the ideological vacuum that followed the Soviet break-up there was a resurgence of all varieties of Islamic belief. Lacking a tradition of centralized Islamic education in Azerbaijan, the country was vulnerable to foreign influence. Islamic influences from abroad—especially from Iran, the Arab countries, and Turkey—quickly established religious communities and supported educational institutions. Iran has created numerous *madrassas* and actively recruits university students, many of whom will be future clerics, to study in Iran. Turkish, Saudi Arabian, Kuwait and Qatar foundations followed a similar path into Azerbaijan as they saw schools and linkages with universities as prime vehicles for expanding their influence.

Today there are 1,750 mosques in Azerbaijan. Since Azerbaijan regained independence, Kuwait built 63 mosques, Turkey 24, Saudi Arabia three, Iran one, and Qatar one. Textbooks for the *madrassas* are provided by the Caucasus Board of Muslims. In order to disseminate other religions in Azerbaijan, some illegal religious communities organize classes after school, mainly by the help of the secondary school teachers (Miri 2009). The activities of religious communities, organizations, and religious/political powers are under the control of the state, which, at the same time, spares no efforts to hinder the formation of the political culture of Islam in the country. The well-respected Baku Islamic University and its four regional branches, controlled by the Caucasus Board of Muslims, are at the center of the fight against illiteracy and the education of men with a "sound and balanced" knowledge of Islam.

In Kyrgyzstan, the use of universities by the Muslim religion to extend soft power and gain competitive advantage is more pronounced. As in Azerbaijan, religion was suppressed under the Soviets. With independence came a surge of Islamic organizations (and, to a lesser degree, Christian organizations) that registered with the Kyrgyz government and were authorized to establish schools and faith-based universities. Islamic organizations were quick to seize the opportunity, establishing *madrassas* which in turn laid the foundation for the seven Islamic institutions of higher education that are currently in Kyrgyzstan (Malikov 2008). Christian organizations, predominately located in Northern Kyrgyzstan, have not established universities, and the main competitors of the Muslim universities are secular private universities such as Kyrgyz-Turkey Manas University, Kyrgyz-Russian Slavonic University, and the American University of Central Asia. In general, the seven Islamic universities in Kyrgyzstan are considered by a substantial portion of the Kyrgyz educated classes to provide a narrow, religiously oriented education inferior to the more innovative international universities. Nevertheless, it is evident that the Islamic universities are a potent force for extending religious and cultural values in the region.

# Conclusion

Centuries ago, Christian missionaries from Europe circled the globe in an effort to convert the "heathen," build cathedrals, and establish their churches. Today, it is the universities which are seen as valuable instruments in efforts to extend political influence and understanding of values, culture and ideals. Universities are inherently adept for this role, as they allow individuals to learn about a nation, religion, language, values, and ideals or culture in a genuine way through immersion and synthesis. Nations and groups will likely continue to cooperate, but also use higher education to compete with their perceived rivals; it seems inevitable that nations will use universities as a primary means to project their soft power and gain strategic political advantage.

How this scenario will play out is far from certain. Questions include the following: How will the increasing use of higher education to project soft power impact faculty and students? Are faculty and administrators the new missionaries? Will the utilization of universities to project soft power in a highly charged, competitive atmosphere ultimately undermine the collaboration necessary to advance knowledge and help address pressing global social, environmental, health and economic problems? Is it an oxymoron to contend that universities can both cooperate and compete in a world where they are increasingly being viewed as instruments for the projection of national, religious or cultural values and agendas?

# References

Adams, Shar. 2007. "'Soft Power' To Be Applied on Campus." *The Epoch Times*, October 15. http://en.epochtimes.com.

Burghart, Dan. 2007. "The New Nomads? The American Military Presence in Central Asia." *China and Eurasia Forum Quarterly* 5 (2): 5–19.

China Central Television. 2009. "Vice President Xi Unveils Confucius Institute on the Caribbean Island." CCTV, February 14. http://www.cctv.com.

China's Ministry of Education. 2008a. *Both the Number of Chinese Students Studying Abroad and Foreign Students Studying in China Surmount 1.2 Million Since China's Opening Up*. Beijing: Ministry of Education. http://www.moe.edu.cn.

———. 2008b. *Foreign Students Coming to China in 2007 Surmount 190,000*. Beijing: Ministry of Education. http://www.moe.edu.cn.

Ding, Sheng, and Robert A. Saunders. 2006. "Talking Up China: An Analysis of China's Rising Cultural Power and Global Promotion of the Chinese Language." *East Asia* 23 (2): 3–33.

Drucker, Peter F. 1993. *Post Capitalist Society*. New York: Harper Collins.

Heyat, Farideh. 2002. *Azeri Women in Transition: Women in Soviet and Post-Soviet Azerbaijan*. London: Routledge.

Joffe, Josef. 2001. "Clinton's World: Purpose, Policy and Weltanschauung." *The Washington Quarterly* 24 (1): 141–154.

Kurlantzick, Joshua. 2007. *Charm Offensive: How China's Soft Power is Transforming the World*. New Haven: Yale University Press.

Lai, Hongyi. 2006. "China's Cultural Diplomacy: Going for Soft Power." East Asia Institute Background Brief No. 308. Singapore: East Asia Institute. http://www.eai.nus.edu.sg.

Lawrence, Dune. 2009. "China Pushes Soft Power." Bloomberg News, February 17. http://www.iht.com/articles.

Lee, Moosung. 2007. "Where Are Global Leaders Educated?" *International Higher Education* 49 (Fall): 6. http://www.bc.edu.

Lum, Thomas, Wayne M. Morrison, and Bruce Vaughan. 2008. *CRS Report for Congress: China's "Soft Power" in Southeast Asia*. Washington, DC: Federation of American Scientists. http://www.fas.org.

Malikov, Kadyr. 2008. *It Is Necessary to Modernize Religious Education in Kyrgyzstan*. Bishkek: 24 News Agency. http://www.24.kg.

Miri, Elshad. 2009. *Religious Communities Financed by External Powers*. Baku: Islam.com.az. http://www.islam.com.az.

National Office for Teaching Chinese as a Foreign Language. 2008. *Location of Confucius Institutes*. Beijing: National Office for Teaching Chinese as a Foreign Language.

National Statistical Committee (Kyrgyz Republic). 2008. *Education and Science in Kyrgyz Republic: The Statistical Collection*. Bishkek: The National Statistical Committee of Kyrgyz Republic.

Nye, Joseph. 2003. "Propaganda Isn't the Way: Soft Power." *International Herald Tribune*, January 10, 6. http://www.nytimes.com.

———. 2005. *Soft Power: The Means to Success in World Politics*. Cambridge, MA: Public Affairs.

———. 2008. *Soft Power and Higher Education*. Boulder, CO: EDUCAUSE. http://net.educause.edu.

Pew Research Center. 2003. *Pew Global Attitudes Project*. Washington, DC: Pew Research Center.

Reychler, Luc. 1997. "Religion and Conflict." *International Journal of Peace Studies* 2 (1): 16.

Stetar, Joseph, Olekisy Panych, and Andrew Tatusko. 2007. "State Power in Legitimating and Regulating Private Higher Education: The Case of Ukraine." In *Private Higher Education in Post Communist Europe*, ed. S. Slantcheva and D. C. Levy. New York: Palgrave Macmillan.

Whitney, Christopher B., and David Schambaugh. 2008. *Soft Power in Asia: Results of a 2008 Multinational Survey of Public Opinion*. Chicago: The Chicago Council on Global Affairs and the East Asia Institute. http://www.thechicagocouncil.org.

Yang, Rui. 2007. "China's Soft Power Projection in Higher Education." *International Higher Education* 46 (Winter): 24. http://www.bc.edu.

Yoshihara, Toshi, and James R. Holmes. 2008. "China's Energy-Driven 'Soft Power.'" *Orbis* 52 (1): 123–137.

# Chapter 15

---

# Internationalization and the Competitiveness Agenda

*Jane Knight*

## Introduction

Internationalization is changing the world of higher education, and globalization is transforming the process of internationalization. It is difficult to imagine another time in history when the pervasive force of globalization has had more impact culturally, economically, and politically. The increased importance of the knowledge enterprise, innovations in information and communication technologies, a stronger orientation to the market economy, growth in regional/international governance systems, and a new emphasis on multilateral trade agreements all contribute to an accelerated flow of people, economy, ideas, culture, technology, goods, and services in our more globalized world. Globalization is neither neutral nor uniform in its impact. Globalization affects countries, cultures, and systems in different ways—some positive, others negative. All sectors of society are being affected; higher education is no exception (Knight 2008a).

Internationalization, a concept which emphasizes relations between and among countries and cultures, differs from globalization, which has at its core the idea of becoming one world—a more interdependent and connected world. In terms of higher education, however, internationalization and globalization are closely linked, with internationalization both reacting to the force of globalization and being an agent of globalization.

This is especially true for cross-border education, which moves people, ideas, knowledge, programs, providers, services, and projects across borders.

Academic mobility has traditionally occurred in the context of exchange and cooperation, but times are changing. While collaborative academic partnerships are still important, the elements of competition, privatization, and commercialization—often seen as globalization practices—are increasingly prevalent in the arena of cross-border higher education, and require closer scrutiny (Currie 2005). The focus of this chapter is the growing importance and impact of this competition and commercial agenda on the internationalization process.

The chapter begins with a discussion on the meaning and use of the term "internationalization." This sets the scene for examining the shift in rationales driving the international dimension of higher education. Opinions of senior academic administrators on the motivations and expectations of internationalization are examined, revealing the link between the current priority being given to competitiveness, and the present preoccupation with international profile, rankings, and being a "world-class" higher education institution (HEI). The following sections briefly address the commercialization and commodification of higher education, the "brain train" phenomenon, regional education hubs, cross border education, and joint/double degree programs, as well as their respective roles in furthering or reacting to the competition agenda. The concluding remarks raise the question of whether increased competitiveness leads to more benefits or risks for higher education, and calls for increased attention to the double role that international higher education plays in furthering cooperation and competition among countries.

The analysis of the opinions of higher education leaders is based on the findings of the Internationalization Survey conducted by the International Association of Universities (IAU 2005). IAU is committed to conducting a worldwide survey of HEIs about new developments and issues related to internationalization every three years. The most recent data included 526 respondents from 95 different countries in all regions of the world. The respondents were, for the most part, senior academic leaders responsible for policy development and oversight of the international work of their university. A major question guiding the data analysis was whether there was a difference in the opinions about internationalization between developing and developed country respondents, or a difference among respondents' views according to the six major regions of the world. Therefore, the findings were analyzed in three ways: (a) aggregate analysis of all respondents; (b) analysis by developing or developed country status (defined by the ranking on the Human Development Index); and (c) analysis by region: Africa, Asia, Europe, Latin America, Middle East, and North America (Knight 2006a).

## Understanding Internationalization

As internationalization changes to meet new challenges, it is important to examine the key concepts that inform and shape the internationalization process, as well as analyze the unexpected developments or consequences related to increasing competitiveness and commodification.

"Internationalization" is a term that means different things to different people. For some people, it means a series of international activities, such as academic mobility for students and teachers, international networks, partnerships and projects, new international academic programs, and research initiatives. For others it means the delivery of education to other countries through new types of arrangements such as branch campuses or franchises, and using a variety of face-to-face or distance techniques. To many, it means the inclusion of an international, intercultural, and/or global dimension into the curriculum and teaching/learning process. Still others see internationalization as a means to improve national or world rankings of their institution, or to recruit the best and brightest of international students and scholars. International development projects have traditionally been perceived as part of internationalization, and more recently the increasing emphasis on trade in higher education is also seen as internationalization. Thus, the concept is used in diverse ways both within and between countries.

Internationalization is not a new term, nor is the debate over its meaning. Internationalization has been used for years in political science and governmental relations, but its popularity in the education sector has soared only since the early 1980s. Prior to this time, "international education" and "international cooperation" were favored terms and still are in some countries. In the 1990s, the discussion centered on differentiating "international education" from such overlapping terms as "comparative education," "global education," and "multi-cultural education." But today, the relationships and nuances of meaning among "cross-border," "transnational," "borderless," and "international" education are more important—and are causing much confusion.

The challenge of developing a definition is the need for it to be generic enough to apply to many different countries, cultures, and education systems. This is no easy task. While it is not the author's intention to develop a universal definition, it is imperative that it can be used in a broad range of contexts and for comparative purposes across countries and regions of the world. With this in mind, we must ensure that a definition does not specify the rationales, benefits, outcomes, actors, activities, and stakeholders of internationalization, as they vary enormously across nations and also from institution to institution. What is critical is that the international dimension relates to all aspects of education and the role that it plays in society.

Internationalization at the national/sectoral/institutional level is defined in this chapter as:

> The process of integrating international, intercultural and global dimensions into the purpose, functions—teaching/learning, research and service—or delivery of higher education. (Knight 2004, 9)

This is intentionally a neutral definition of internationalization. Many would argue that the process of internationalization should be described in terms of promoting cooperation and solidarity among nations, improving quality and relevance of higher education, or contributing to the advancement of research for international issues. While these are noble intentions, and internationalization can contribute to these goals, a definition needs to be objective enough that it can be used to describe a phenomenon which is in fact universal, but which has different purposes and outcomes, depending on the actor or stakeholder.

A significant development in the conceptualization of internationalization in the last five years has been the introduction of the terms "internationalization at home" and "cross-border education." Campus-based strategies are most often referred to as internationalization "at home," and off-campus initiatives are called "cross-border education." As a result of a heightened emphasis on international academic mobility, the "at home" concept has been developed to give greater prominence to campus-based elements, such as the intercultural and international dimension in the teaching/learning process, research, extra-curricular activities, relationships with local cultural and ethnic community groups, and the integration of foreign students and scholars into campus life and activities.

It is important to point out that the internationalization process consists of these two separate but closely linked and interdependent pillars. Cross-border education has significant implications for campus-based internationalization and vice versa. Interestingly enough, many of the new developments and the growing competitiveness are associated with the cross-border aspects of internationalization—and thus, cross-border education such as branch campuses, franchises, and joint/double degrees are the primary focus of this chapter.

# Rationales: A Shift in the Motivations and Expectations of Internationalization

The motivations driving internationalization are core to understanding all aspects of the international dimension of higher education. They help to explain why an institution or a country believes internationalization is important, what strategies are used, which benefits are expected, and which risks are taken or feared. At a more fundamental level, rationales reflect the core values that a higher education system holds regarding the contribution that international, intercultural, and global elements make to the role of higher education in society.

The rationales underpinning the process of internationalization have been changing in the last two decades. Much has been written about the shift from academic and social/cultural rationales to economic and political rationales (de Wit 2002; Altbach and Knight 2007; Dunn and Nilan 2007). But the increasing emphasis on competition over cooperation, and the reality of international cooperation for the sake of unilateral competitiveness, are only just beginning to be explored more seriously.

In the IAU worldwide survey on internationalization, respondents were asked to identify the top rationales driving internationalization at the institutional level and at the national/country level. Rationales at the national level are most relevant to this discussion, as the issues of competition and cooperation were directly addressed. Respondents were asked to rank the top three rationales for internationalization from the following list: (a) increase competitiveness (scientific, technological, economic); (b) promote international solidarity and cooperation; (c) develop strategic alliances (political, cultural, academic, trade); (d) strengthen educational export industry; (e) build a country's resource capacity; (f) further cultural awareness and understanding; and (g) contribute to regional priorities and integration.

The survey results indicated that the number one perceived rationale driving countries to internationalize their higher education system is "to increase competitiveness-scientific, technological and economic." This is followed by the second top ranking rationale—"to develop strategic alliances-political, cultural, academic, trade"—which, in reality, is closely linked to the competitiveness rationale. It is revealing that to "promote international cooperation and solidarity" was ranked fourth. In fact, twice as many respondents ranked "competitiveness" as the number one rationale, compared to those who ranked "international cooperation" as the number one priority. This dramatically illustrates the current perceived priority given to competition. "Cultural awareness," which one could say is closely linked to cooperation, ranked fifth, followed by "strengthen exports," and lastly, "regional priorities."

An interesting question is whether there is a difference between developing and developed countries with respect to the priority given to rationales. The survey findings indicate that the answer is no. This is contrary to the expectation that developing countries would have ranked international cooperation and solidarity higher, given their past and current involvement in development cooperation projects. Post–World War II, the international work of HEIs involved a great deal of development cooperation work through scholarship programs, technical assistance, and capacity building projects. In the last two decades there has been a gradual and discernible shift from development cooperation to academic partnerships, and more recently to commercial competition. This is often labeled the "aid to trade" shift (Knight 2007).

Did the IAU survey data show a difference in orientation to competitiveness and cooperation rationales among the six regions of the world? The answer is definitely yes. Europe and Latin America stand out with the largest number of respondents ranking competitiveness as their first priority while Africa had the least number. Of greatest interest is the importance that European respondents attach to competitiveness. One could say that this is in line with the policies and intentions of the Bologna reform process (Tauch 2005; de Wit 2006), which has identified "increased attractiveness" (often interpreted as code for "increased competitiveness") as a major goal. The survey findings support the perception that European universities want to be seen as competitive, especially in relation to the US. Furthermore, it appears that Europe is successful in this regard, given that Europe ranked highest as the region of the world with whom universities would like to work—that is, after intra-regional collaboration, which ranked first for all regions except North America and the Middle East.

The ranking of institutional level rationales did not specifically address competitiveness and cooperation, but nevertheless, it does shed some light on this issue. The two rationales that respondents identified as being most important at the institutional level were "increase student and faculty international knowledge and intercultural understanding" and "strengthen research knowledge capacity and production." Both of these rationales speak to the core mission of universities—teaching/learning, research, and service. While it appears that internationalization, first and foremost, should be contributing to the core purpose and activities of their institutions, there is nonetheless some convergence between the competitiveness rationale and the strengthening of research knowledge capacity and production, given the dominance of the knowledge economy.

The third-ranking rationale at the institutional level is to "create international profile and reputation." In fact, "international profile" ranked higher than "improvement of academic quality," which is revealing. Using internationalization as a means to create an international "brand" is closely linked to increasing competitiveness, and appears to be seen as more important than improving academic quality. Of course, the most critical question here is the link between international rankings and quality: does an international profile or high ranking actually mean high quality? There is no consensus on this point.

## Status and Profile: World Rankings

There is no question that international and regional rankings of universities have become more popular and problematic in the last five years (Horn et al. 2007; Marginson and van der Wende 2007). The heated debate about their validity, reliability, and value continues. But at the same time, university presidents declare in their strategic plan that a measurable outcome of internationalization will be the achievement of a specific position in one or more of the global ranking instruments. As the IAU survey results indicated, internationalization is perceived by some institutions as a means to gaining worldwide profile and prestige.

The intense competition for world rankings would have been impossible to imagine a mere twenty years ago, when international collaboration among universities, through academic exchanges and development cooperation projects, was the norm. Of course, these types of activities still take place, but the factors influencing internationalization are becoming increasingly varied, complex, and competitive. Is international cooperation and exchange becoming overshadowed by competition for status, bright students, talented faculty, research grants, and membership in networks? The answer is probably yes, especially for those universities already placed in the top 25 percent of their national league tables, or the two international ranking systems developed by the Shanghai Jiao Tong University Institute of Higher Education and *Times Higher Education*. Asia is one region of the world which takes competitiveness and ranking very seriously (Mok 2007). National programs, such as Projects 211 and 985 in China, or the Brain Korea 21, aim to help their best research universities improve their international rankings through increased funding programs (Kim and Nam 2007; Liu 2007).

The issue of "world-class universities" is a subject of intense scrutiny (Altbach and Balan 2007), but "world-class" is still in the eye of the beholder, in spite of all the attention being given to rankings. The presence of national, regional, international, and discipline/profession-specific rankings allows universities of all types to be deemed "prestigious" by some self-appointed ranking body, whether it be magazines, consumer guides, universities, or private companies. But, in spite of what is said about the "hollowness" of the ranking game, the competition to be ranked as a "world-class" institutions is increasing, not diminishing.

## Commercialization: A Risk or Benefit?

While the process of internationalization affords many benefits to higher education, it is clear that there are serious risks associated with this complex and growing phenomenon. According to the results of the IAU survey, there is overwhelming agreement (96 percent of respondents) that internationalization brings benefits to higher education. Yet, this consensus is qualified by the fact that 70 percent also believe there are substantial risks associated with the international dimension of higher education.

Overall, the number one risk identified in the survey was the "commodification and commercialization" of education programs. Of interest is that both developing and developed countries identified commercialization as the number one risk—a strong testimony to its importance. A regional level analysis showed that four regions (Africa, Asia Pacific, Europe and North America) ranked commercialization as the top risk, whereas Latin America placed "brain drain" as number one and the Middle East ranked "loss of cultural identity" in first place.

The General Agreement on Trade in Services (GATS) has been a wake-up call for higher education around the world. Higher education has traditionally been seen as a "public good" and a "social responsibility." But, with the advent of this new international trade agreement, higher education has become a tradable commodity, or more precisely in GATS terms, an internationally tradable service (Knight 2006b). Many see GATS as presenting new opportunities and benefits, while others see it as introducing new risks. In addition, there are many who question why the trade sector needs to impose regulations at all, given that the education sector has been using its own agreements and conventions for decades.

At the heart of the debate for many educators is the impact of increased commercial cross-border education on the purpose, role, and values of higher education. The growth in new commercial and private providers, the commodification and market orientation of education, and the prospect of new trade policy frameworks are catalysts for stimulating serious reflection on the role, social commitment, and funding of public HEIs in society. The trinity of teaching/learning, research, and service to society has traditionally guided the evolution of universities and their contribution to the social, cultural, human, scientific, and economic development of a nation. Is the combination of these roles still valid, or can they be disaggregated and rendered by different providers?

# Competition and the Brain Train

Little did we know twenty five years ago that the highly valued and promoted notion of "international academic mobility" for students, scholars and professors would have the potential to grow into a highly competitive international recruitment business. Several countries are investing in major marketing campaigns to attract the best and brightest talent to study and work in their institutions in order to supply the heavy demand for human resources for innovation and research agendas (Agarawal et al. 2007). The complexities and challenges related to the phenomenon of academic and professional mobility should not be underestimated; nor should the potential benefits. But it is impossible to ignore the latest race for attracting international students and academics for "brain power" and for "income generation." The original goal of cooperating to help students from developing countries move to another country to complete a degree and return home is fading fast, as nations compete for attracting and retaining brain power.

While "brain drain" and "brain gain" are well known concepts (Crush et al. 2006), research is showing that international students and researchers are increasingly interested in taking a degree in country A, followed by a second degree or perhaps internship in country B, leading to employment in country C and probably D, finally returning to their home country after eight to 12 years of international study and work experience. Hence, the emergence of the term "brain train" (Knight 2008c). In the final analysis, whether one is dealing with brain gain, brain drain, or brain train, this phenomenon is presenting benefits, risks and new challenges for both sending and receiving countries.

From a policy perspective, higher education is becoming a more important actor, and is now working in closer collaboration with immigration, industry, and the science and technology sectors to build an integrated strategy for attracting and retaining knowledge workers. The convergence of an aging society, lower birth rates, the knowledge economy, and professional labor mobility is introducing new issues and opportunities for the higher education subsector while at the same time, encouraging unprecedented competition for recruiting the best and the brightest students and scholars.

# Cross-Border Education: Student Recruitment to Program Franchises to Education Cities

The business of cross-border education is booming in response to the escalating demand for a skilled workforce, a pronounced orientation to a market economy, the commercialization of education as a trade commodity, and the escalating numbers of students wanting higher education opportunities and a foreign credential (Vincent-Lancrin 2004; Sackman 2007). Academic mobility between countries is nothing new to higher education; it has been happening for centuries. But the growth in cross-border education such as branch campuses, franchising, and virtual education demonstrates that it is no longer students and scholars who move; programs and higher education providers also move. Furthermore, countries propelled by the need to establish a firm footing in the knowledge economy are moving to second generation cross-border education strategies: regional education hubs, economic free zones, education cities, knowledge villages, gateway, and hot spots.

These initiatives include foreign universities establishing branch campuses or offering education and training programs to local and international students. Financial incentives are often offered through favorable tax schemes or land and infrastructure arrangements. An important feature is co-locating a critical mass of foreign universities and students with private companies, research and development enterprises, and science and technology parks to collectively support and develop new knowledge industries. No matter how distasteful academics may find the treatment of education as a commodity, the development of these education hubs/cities are positive proof that education is often treated as a commodity to be acquired and used to gain competitive advantage in the knowledge economy.

# Double and Joint Degree Programs: Is There a Link to Competition?

Improvement in the quality of research, the teaching/learning process, and curriculum has long been heralded as a positive outcome of international collaboration. Through exchange of good practice, shared curricular reform, close research cooperation, and mobility of professors and students, there is much to be gained through internationalization. A recent trend has been the establishment of joint programs between institutions in different countries that lead to double (or multiple) degrees, and in some cases a joint degree—although the latter faces steep legal constraints (EUA 2004). Joint programs are intended to provide a rich international and comparative academic experience for students, and to improve their opportunities for employment. But, with all new ideas come questionable adaptations and unintended consequences. For instance, in some cases, double degrees can be nothing more than double counting one set of course credits. Situations exist where two or three credentials (one from each participating institution) are conferred for little more than the workload required for one degree.

While it may be very attractive for students to have two degrees from two institutions in two different countries, the situation can be described as the thin edge of academic fraud if course requirements for two full degrees are not completed or differentiated learning outcomes are not achieved (Knight 2008d). It is important to point out that there are many excellent and innovative joint and double degree programs being offered—especially by European institutions, given the priority they have in the Bologna Process (Schule 2006). When one asks the question of why and how these new models for double/joint degree programs have evolved in the last five years, there is a wide range of responses, and competition is one of the factors. But, in this scenario, competition can either be the stimulus for innovation and the development of new approaches to joint programming models, or it can be seen as a factor leading to misrepresentation of the actual coursework completed.

## Concluding Remarks

As discussed, internationalization of higher education was originally conceived in terms of exchange and sharing of ideas, cultures, knowledge, and values. Formalized academic relations between countries were normally expressed in bilateral cultural and scientific agreements. Today, the agreements are often trade-related, economic, and/or political, thus showing a significant shift from the original idea of academic exchange to increased economic motives and commercial competition.

The explosion of the knowledge economy in this era of globalization is escalating competition between nations and regions and increasing the importance of higher education as a political and economic actor. As a result, the international dimension of higher education is experiencing a fundamental shift in its approach, values, and strategies. The traditional hallmarks of international academic relations—collaboration; and the cooperation and exchange of students, scholars, and research—have been eroded in the past decade in favor of competition, commercialization and trade, as increasingly important elements and drivers of internationalization (Knight 2008b).

There is no doubt that internationalization of higher education is being fundamentally changed in reaction to and support of the competition agenda and market orientation. But whether this change has more positive or negative consequences is yet to be determined. What is certain is that it brings new opportunities, risks, benefits, and challenges. While the tenor of this discussion has leaned toward caution about where the competitiveness agenda is heading, it is imperative that serious attention be given to the new opportunities and innovation that competition can offer. To not take advantage of the benefits of competition, or to be blind to its unintended and negative consequences, is equally problematic. In short, the double role of internationalization in furthering both cooperation and competition among countries is a new reality of our more globalized world, and the delicate balance between these two roles warrants further research and attention.

# References

Agarwal, Pawan, Mohsen Elmahdy Said, Molatlhegi T. Sehoole, Muhammad Sirozi, and Hans de Wit. 2007. "The Dynamics of International Student Circulation in a Global Context." In *Higher Education in the New Century, Global Challenges and Innovative Ideas*, ed. P. Altbach and P. McGill Peterson. Rotterdam: Sense Publishers.

Altbach, Philip, and Jorge Balan, eds. 2007. *World Class Worldwide: Transforming Research Universities in Asia and Latin America*. Baltimore, MD: Johns Hopkins University Press.

Altbach, Philip, and Jane Knight. 2007. "The Internationalization of Higher Education: Motivations and Realities." *Journal of Studies in International Education* 11 (3/4): 290–305.

Crush, Jonathan, Wade Pendelton, and Daniel S. Tevera. 2006. "Degrees of Uncertainty; Students and the Brain Drain in Southern Africa." In *The Internationalization of Higher Education in South Africa*, ed. R. Kishun. Durban: International Education Association of South Africa.

Currie, Jan. 2005. "Privatization and Commercialization: Two Globalizing Practices Affecting Australian Universities." In *Globalization and Higher*

*Education,* ed. A. Arimoto, F. Huang, and K. Yokoyama. Higashi-Hiroshima, Japan: Research Institute for Higher Education, Hiroshima University.

de Wit, Hans. 2002. *Internationalization of Higher Education in the United States of America and Europe: A Historical, Comparative, and Conceptual Analysis.* Westport, CT: Greenwood Publishers.

———. 2006. "European Integration in Higher Education: The Bologna Process Towards a European Higher Education Area." In *International Handbook of Higher Education, Part Two,* ed. J. Forest and P. Altbach. Dordecht, The Netherlands: Springer.

Dunn, Mel, and Pamela Nilan. 2007. "Balancing Economic and Other Discourses in the Internationalization of Higher Education in South Africa." *International Review of Education* 53 (3): 265–281.

European Union Association. 2004. *Developing Joint Masters Programmes for Europe.* Brussels: European University Association.

Horn, Aaron S., Darwin D. Hendel, and Gerald W. Fry. 2007. "Ranking the International Dimension of Top Research Universities in the United States." *Journal of Studies in International Education* 11 (3–4): 330–358.

International Association of Universities. 2005. *IAU 2005 Internationalization Survey: Preliminary Findings Report.* Paris: United Nations Educational, Scientific and Cultural Organization. http://www.unesco.org.

Kim, Ki-Seok, and Sughee Nam. 2007. The Making of a World-Class University at the Periphery: Seoul National University. In *World Class Worldwide: Transforming Research Universities in Asia and Latin America,* ed. P. Altbach and J. Balan. Baltimore, MD: Johns Hopkins University Press.

Knight, Jane. 2004. "Internationalization Remodeled: Rationales, Strategies and Approaches." *Journal for Studies in International Education* 8 (1): 5–31.

———. 2006a. *Internationalization of Higher Education: New Directions, New Challenges. 2005 IAU Global Survey Report.* Paris: International Association of Universities. http://www.unesco.org.

———. 2006b. *Higher Education Crossing Borders: A Guide to the Implications of GATS for Cross-Border Education.* Paris: Commonwealth of Learning and United Nations Educational, Scientific and Cultural Organisation. http://unesdoc.unesco.org.

———. 2007. "Cross-Border Tertiary Education: An Introduction." In *Cross-Border Tertiary Education: A Way Towards Capacity Development.* Paris, Washington, DC and The Hague: Organisation for Economic Co-operation and Development, World Bank, and Netherlands Organization for Cooperation in International Higher Education.

———. 2008a. *Higher Education in Turmoil: The Changing World of Internationalization.* Rotterdam: Sense Publisher.

———. 2008b. "The Role of Cross-border Education in the Debate on Education as a Public Good and Private Commodity." *Journal of Asian Public Policy* 1 (2): 174–188.

———. 2008c. "The Internationalization of Higher Education: Are We on the Right Track?" *Academic Matters: The Journal of Higher Education.* October/November: 5–9.

Knight, Jane. 2008d. *Double and Joint Degrees: Vexing Questions and Issues.* London: Observatory on Borderless Higher Education. http://www.eahep.org.

Liu, Nian Cai. 2007. "Research Universities in China: Differentiation, Classification and Future World-class Status." In *World Class Worldwide: Transforming Research Universities in Asia and Latin America*, ed. P. Altbach, and J. Balan. Baltimore: The Johns Hopkins University Press.

Marginson, Simon, and Marijk van der Wende. 2007. "To Rank or To Be Ranked: The Impact of Global Rankings in Higher Education." *Journal of Studies in International Education* 11 (3–4): 306–329.

Mok, Ka Ho. 2007. "Questing for Internationalization of Universities in Asia: Critical Reflections." *Journal of Studies in International Education* 11 (3–4): 433–454.

Sackmann, Reinhold. 2007. "Internationalization of Markets for Education: New Actors Within Nations and Increasing Flows Between Nations." In *New Arenas of Education Governance*, ed. K. Martens, A. Rusconi, and K. Leuze. New York: Palgrave Macmillan.

Schule, Ulrich. 2006. *Joint and Double Degrees within the European Higher Education AreaóTowards Further Internationalization of Business Degrees.* Papers on International Business Education. Paris: Consortium of International Double Degrees.

Tauch, Christian. (2005). "The Bologna Process: State of Implementation and External Dimension." In *Opening Up to the Wider World: The External Dimension of the Bologna Process*, ed. F. Muche. Bonn: Lemmens.

Vincent-Lancrin, Stéphan. 2004. "Implications of Recent Developments for Access and Equity, Cost and Funding, Quality and Capacity Building." In *Internationalisation and Trade in Higher Education*. Paris: Centre for Educational Research and Innovation and OECD.

# Chapter 16

# Global Competition in Higher Education: A Comparative Study of Policies, Rationales, and Practices in Australia and Europe

*Hans de Wit and Tony Adams*

## Introduction

Over the past two decades, competition has become a central preoccupation in higher education, and has moved from the national to the regional and international arenas. The global knowledge economy not only has forced higher education to respond to this development but also has stimulated nations and institutions to become important actors and competitors. It is, however, an overestimation to argue that this is true for all institutions of higher education in all countries and continents in the same way and at the same time.

A comparison of Australia and Europe illustrates the diverse ways higher education responds to an increasingly international competitive environment. The higher education subsector in Australia and the United Kingdom had, by the mid-1980s, shifted from aid to trade in their international orientation. In continental Europe, this shift has been less radical, taken more time, and occurred via a shift from aid to cooperation and exchange first, before moving toward competition.

In this article we explain the rationale behind these different approaches, and the main trends, opportunities, and risks that are present. We analyze the development of international competition in higher education in the two continents within the context of the worldwide changing dynamic in internationalization of higher education.

## Globalization and Internationalization

The landscape of international higher education has been changing over the past 15 years (de Wit 2002, 2008a; Knight 2008). The international dimension and the position of higher education in the global arena are more dominant than ever in international, national, and institutional documents and mission statements. Higher education is increasingly influenced by globalization but is also becoming a more dynamic actor in the global knowledge economy. Globalization and the role of higher education in it are linked by: (a) an increasingly unmet demand for higher education in the world; (b) growth in the number and types of new "for-profit" providers in addition to public universities; (c) "not-for-profit" private universities; and (d) the emergence of new, innovative, cross-border delivery.

Ulrich Teichler (2004), Peter Scott (2005), Philip Altbach (2006), Hans de Wit (2008a), and Jane Knight (2008), and others address the complex relationship between globalization and internationalization of higher education. According to Scott (2005) "the distinction between internationalisation and globalisation, although suggestive, cannot be regarded as categorical. They overlap, and are intertwined, in all kinds of ways" (14). Altbach (2006) defines globalization as "the broad economic, technological, and scientific trends that directly affect higher education and are largely inevitable in the contemporary world," where internationalization "refers to specific policies and programs undertaken by governments, academic systems and institutions, and even individual departments to support student or faculty exchanges, encourage collaborative research overseas, set up joint teaching programs in other countries, or a myriad of initiatives" (123).

Teichler (2004) states that "globalisation initially seemed to be defined as the totality of substantial changes in the context and inner life of higher education, related to growing interrelationships between different parts of the world whereby national borders are blurred or even seem to vanish" (23). But, according to Teichler, in recent years the term "globalization" has been substituted for "internationalization" in the public debate on higher education, resulting at the same time in a shift of meanings; he notes that "the term tends to be used for any supra-regional phenomenon rela higher education and/or anything on a global scale related to higher educa tion characterised by market and competition" (24). Teichler defines inter-nationalization as "the totality of substantial changes in the context and inner life of higher education relative to an increasing frequency of border-crossing activities amidst a persistence of national systems, even though some sign of 'denationalisation' might be observed" (22–23).

Frans van Vught et al. (2002), meanwhile, state that "in terms of both practice and perceptions, internationalization is closer to the well-estab-lished tradition of international cooperation and mobility and to the core values of quality and excellence, whereas globalization refers more to com-petition, pushing the concept of higher education as a tradable commodity and challenging the concept of higher education as a public good" (17).

For Jane Knight (2008),

> *globalization* is the process that is increasing the flow of people, culture, ideas, values, knowledge, technology, and economy across borders, result-ing in a more interconnected and interdependent world. Globalization affects each country in different ways and can have positive and/or nega-tive consequences, according to a nation's specific history, traditions, cul-ture, priorities, and resources. Education is one of the sectors impacted by globalization. (xi)

*Internationalization* for Knight is "the process of integrating an interna-tional, intercultural or global dimension into the purpose, functions or delivery of post-secondary education" (xi). Even though different accents are made, one can say that globalization is a social, economic, and polit-ical process to which higher education responds, and in which it is an actor, while internationalization is the way higher education responds to and acts in it.

According to Teichler (2004) there is a growing emphasis within higher education on marketization, competition, and management, some-thing also stressed by others. Bob Reinalda and Ewa Kulesza (2005), for instance, note that

> since the end of the last century, a shift in higher education has taken place from the public to the private domain, parallel to an increase in interna-tional trade in education services.... These developments enhance the sig-nificance of the education market as an international institution, but also contribute to changing the structure of that market. In doing so, an increase in worldwide competition is being revealed. (99)

Several authors call for more attention to social cohesion and to the pub-lic role of higher education as an alternative force to its growing emphasis on competition, markets and entrepreneurialism. Rajani Naidoo and Ian Jamieson (2005) state that "the forces unleashed on higher education in the present context have propelled universities to function less as institutions with social, cultural and indeed intellectual objectives and more as producers of commodities that can be sold in the international marketplace" (39). These concerns also came to the surface in the response of higher education organizations around the world, to the inclusion of education in the General Agreement on Trade and Services (GATS) of the World Trade Organization (WTO).

Notwithstanding these concerns, internationalization of higher education is influenced by the global knowledge economy, and is moving from a cooperative to a more competitive approach. Knight (2008) concludes that the following two components are evolving: (a) *internationalization at home*: activities that help students develop international understanding and intercultural skills (curriculum-oriented) and that prepare students to be active in a much more globalized world; (b) *internationalization abroad*: all forms of education across borders, including circulation of students, faculty, scholars, and programs.

Ultimately, a competitive higher education subsector requires a strong focus "at home" and "abroad." The following comparison of Europe and Australia illustrates that the increasing attention to these two components and the strong interconnection between them, as well as the increasingly competitive character of internationalization in higher education.

# Internationalization and Competition in Europe

Higher education in Europe in the first decade after the Second World War was not very international. The focus was on the reconstruction of its countries after the Great Depression followed by the impact of the Second World War on society and economy. What little international dimension existed was primarily the circulation of elite degree-seeking students in developing countries to the colonial and imperialist powers they were linked to: the United Kingdom, France, Germany, and, to a lesser extent, countries like Belgium and the Netherlands. In addition, governments signed cultural and scientific agreements to exchange small numbers of students and staff.

In the 1960s, another international dimension in higher education emerged: technical assistance, or development aid. The changing relationships between the former colonial powers and the developing world required a different approach. In addition to the traditional circulation of the elites, scholarship schemes provided wider opportunities for students from developing countries to study in Europe, primarily in the countries with which they had traditional cultural and linguistic ties (such as Germany, France, and the United Kingdom; which up until now have continued to be the main receivers of international students, after the United States) and/or political links (such as the USSR). At the same time, capacity and institution building programs offered academic expertise and material support to the higher education subsector in developing countries. This trend was quite widespread, though most noticeable in Scandinavia, the Netherlands, the United Kingdom, and Germany.

The international dimension of European higher education in the 1960s and 1970s was still marginal, and dominated primarily by the circulation of students from developing countries to Europe, some outward circulation of students and scholars from Europe to the United States, and by development aid. In the 1980s, two different shifts occurred in Western Europe. The "benevolent laisser-faire" policy (Barron 1993, 50) and "humanitarianism and internationalism" (Chandler 1989, viii) that characterized the previous decades did not completely disappear, but were bypassed by new policies. In continental Europe a shift took place toward more controlled reception of degree-seeking international students and toward cooperation and exchange (i.e., student and staff mobility), while in the United Kingdom there was a shift to active recruitment of fee-paying international students.

The decision in 1979 by the British government to introduce full-cost fees for foreign students (a move from "aid-to-trade") resulted in a more competitive higher education subsector. In continental Europe the introduction of full-cost fees and higher education as an export commodity remained anathema. On the continent a different move took place, from aid to cooperation and exchange. Under the impetus of the European Commission, programs were designed to stimulate cooperation in research and development (R&D), and in education. From the early 1970s, in Sweden and Germany, and later elsewhere, programs were developed to stimulate cooperation and exchange; most countries had international academic agreements and were involved in the Fulbright Program with the United States.

During the 1970s, the European Commission started to stimulate R&D cooperation, and in 1976 introduced a pilot program known as the "Joint Study Programmes Scheme" to stimulate academic mobility, but the impact of these programs was marginal. In the 1980s, these initiatives at the national and European level contributed to the creation of the so-called Framework Programs for R&D in 1984, and the European Action Scheme for the Mobility of University Students (ERASMUS) in 1987.

The driving rationales behind these initiatives were both Europeanization and strengthening Europe's position in the global economy.

Although the United Kingdom, as a member of the European Union, was involved in these developments, its participation in the educational programs has been marginal. There was, and remains, a tension between the more competitive approach to recruitment of fee-paying students (a focus on degree-seeking student circulation) and the subsidized programs of the European Commission, based on the principle of exchange (a focus on mobility as part of the home degree). The reputation of British higher education, its extended network of Commonwealth countries, the dominance of English as a first or second language, and the financial necessity to recruit full-cost students from abroad placed British higher education in a position to be a competitive player in the international student market, as well as in the cross-border delivery of education, just behind the United States.

By the end of the 1990s, first in the Netherlands and Scandinavia and later in Germany and France, a shift to higher education as an export commodity began to emerge. Although several countries—Denmark, the Netherlands, Sweden, and Finland—have or are planning to introduce full-cost fees for non-EU international students, the main drive has not been income generation, as was the case in the United Kingdom.

Most of Europe, in particular the larger economies of Germany, France, Italy, Spain, and Scandinavia, have zero or low tuition fees for domestic and international students. At the national and European level, the drivers have been increasing competitiveness of higher education in the global knowledge economy, and the establishment of a national/European brand and status of higher education, society, and economy.

More recently, global competition for highly skilled manpower has become a strong pull factor in international student circulation. The graying societies of Europe compete globally for top talent to fill the gaps in their knowledge economies. Migration and circulation of the highly skilled, and global competition for talent are terms that are becoming more dominant rationales. At the institutional level, rationales such as international classrooms, intercultural and global competencies, recruitment of top talent students and scholars, and institutional profile and status, are setting the scene.

In 2002–2003, there were 1.1 million foreign students enrolled in higher education in the so-called EURODATA region (comprised of the 27 EU nations; the four European Free Trade Agreement members of Switzerland, Iceland, Lichtenstein, and Norway; as well as Turkey). Of these, 46 percent are nationals from within this group of 32 countries, while 54 percent are from outside. More than 60 percent of these students study in the three main countries: the United Kingdom, Germany, and France. France is a different destination country than the others, as only 28 percent of students are European and 51 percent are African students. As far as outward mobility is concerned, only 575,000 students, or 3 percent were studying abroad in 2002–2003, of which 81 percent were in another EURODATA country and 13 percent were in the United States (de Wit 2008b, 184–193).

Cross-border delivery of education is a major and growing market for the United Kingdom but is still marginal in continental Europe. But, as a destination market, Central and Eastern Europe as well as Southern Europe are experiencing an increasing presence of foreign programs and providers.

The Bologna Declaration of 1999 and the Lisbon Agenda of 2001 are manifestations of the need, and joint efforts by governments, the private sector, and higher education to reform higher education in Europe toward becoming more competitive in the global knowledge economy. Van der Wende (2001) speaks of a change in paradigms from cooperation to competition, although as she writes: "Not surprisingly most continental European countries pursue a cooperative approach to internationalization, which in terms of international learning and experience is more compatible with the traditional value of academia" (255). In a benchmarking exercise about the internationalization strategies of five European universities, de Wit (2005) encountered clear differentiations between universities, in particular universities from the United Kingdom, Northern Europe, and Southern Europe.

Although there is an increasing emphasis on economic rationales and competition, the changing landscape of internationalization is not necessarily developing in similar ways everywhere in Europe. Internationalization strategies are filtered and contextualized by the specific internal context of the university and its national embeddedness (Frolich and Veiga 2005). The recent emphasis on competition for talents, as well as the reforms undertaken by the Bologna Process, have brought continental Europe and the United Kingdom closer in their approaches. At the same time, the United Kingdom and the rest of Europe are increasingly concerned by the coordinated approach to international student recruitment and cross-border delivery of education in Australia.

## Internationalization and Competition in Australia

In January 1950, foreign ministers representing seven Commonwealth countries (Australia, Canada, Ceylon, India, New Zealand, Pakistan, and the United Kingdom), met to form the Colombo Plan. The plan was a cooperative venture for the economic and social advancement of the peoples of South and Southeast Asia, leading to international students coming to Australia as part of that country's bilateral aid program. Prior to this period, international students were regulated by Australia's then racially based immigration policy, with country-by-country concessions to allow students to enter Australia (Back and Davis 1995, 123). It is estimated that prior to World War II there were no more than 500 international students enrolled in Australian universities.

In 1974, the Australian government abolished university fees, including those for international students, and established a quota of 10,000 international students at any given time. The so-called "White Australia policy" had also been abandoned. The quota was replaced in 1979 in favor of unofficial country-by-country quotas and an Overseas Student Charge levy amounting to 10 percent of the real cost of an international student's education (Back and Davis 1995, 123). This percentage was increased over time, with the balance of the student's fee coming from the Australian aid budget.

By the mid 1980s, this first phase of Australian international education could be described as aid-based (sponsored students) with private international students paying up to a third of the notional real fee (subsidized students) and the remainder of the fee coming via the aid program. There was little emphasis by institutions on student mobility and developing international links. During this period, many aspiring academics and researchers undertook higher degree studies overseas, often in the United Kingdom, United States, or Canada.

In 1984, the Australian government received the report of the Committee to Review the Australian Overseas Aid Program, which argued that education was an export industry within which universities could compete for students and funds. This fundamental change—accepted by the government, and signaling a second, commercially based phase—caused anguish among academics, administrators, the community, and diplomats, with criticism coming to Australia mainly from Southeast Asia (Back and Davis 1995; Cuthbert, Smith, and Boey 2008).

A minimum fee was set by the Australian government for each discipline, and rules were established to ensure that international students were not cross subsidized by the taxpayer. Australia continued to provide significant numbers of scholarships within the Asia Pacific Region through the bilateral aid program. International student numbers grew from 16,782 in 1986, to 34,401 in 1991, as universities established the infrastructure to recruit and support international students (Back and Davis 1995, 127). In some cases, the entire international dimension of the university was owned and controlled by the university.

International offices were created with a range of commercial and non-commercial activities, but the focus was firmly on the recruitment of fee-paying international students, and for providing services to those students. This infrastructure gradually became the basis for a wider interpretation of international education. Tony Adams (1998) notes that by 1991–1992, Australian universities "had begun to seriously internationalise other aspects of university life" (5). The need to internationalize the experience of students and staff, to internationalize the curriculum, and to support international students together began to be seen as increasingly important, for reasons similar to those in Europe.

There was no "roadmap" to guide international staff on how to recruit international students; administrators and academics learned the skills as they went. An entrepreneurial spirit prevailed, with the appointment of agents, the establishment of academic pathways and English-language preparation programs, the creation of tertiary-level diplomas and transnational programs, and the addressing of pastoral and academic needs of incoming students. Dennis Blight, former CEO of the International Development Program (IDP) of Australian Universities noted that Australian Universities are vigorously competitive. Australia's color and splash in the market place has been important to success but not the keys to its competitiveness. The key factors are business like attitudes, and a willingness to invest in market and product development (Blight 1998, cited in Adams 2004).

This all contributed to a vibrant export scenario in universities and "technical and further education" (TAFE) colleges, as well as the rapidly growing private sector, but mistakes were made, and the Australian government began to see risks to Australia's reputation. Following the collapse of a number of private colleges, and concerns over the practices of unscrupulous and naive operators, the Australian government introduced the *Educational Services for Overseas Students (ESOS) Act* to regulate the activities of both private and public educational providers.

The Act, now in its third iteration, is a powerful piece of consumer protection legislation that deals with the marketing, recruiting, teaching, and supporting of international students. Australian Education International (AEI), a branch of the Australian Ministry of Education, administers the Act and provides generic marketing and government-to-government services through a series of offices globally. AEI also provides a comprehensive range of recruitment statistics, both in the public domain and for subscribing educational providers (see www.aei.gov.au), and has carried out major branding studies around Australia's value proposition. These studies have consistently identified lifestyle, safety, proximity to Asia, and "value-for-money" as key elements of Australia's "brand."

The development of an Australian government interest and policy framework that existed in no other country was the beginning of a public-private partnership between national and state governments, public and private educational institutions, professional and industry associations, and commercial stakeholders (such as agents and health-care providers). This both facilitates the activity and protects educational standards and student consumers, and has been a major contributor to Australia's success.

Australia's Department of Immigration and Citizenship (DIAC) developed and published country-by-country student visa requirements through a five level assessment framework (ADIAC 2008). DIAC also administers the Skilled Migration Program. A controversial "outcome" of the International Student Program is that international students contribute to Australia's skilled migration needs.

Transnational programs grew in parallel with international student recruitment to Australia. These were generally established by entrepreneurial individuals and departments, without capital injection from the home university, and without external funding. They consisted of an Australian university and a host partner (often a professional association or private college with access to physical facilities) twinned together to operate all or part of a degree program in the host country (Adams 1998).

During the 1990s, universities moved to strategize these programs and to introduce strict approval processes, as well as significant quality assurance mechanisms. The Australian Universities Quality Agency (AUQA), which carries out five-year audits of university academic operations, has—because of consistent criticisms of the quality of transnational programs—focused a significant portion of its activities on them. Since AUQA findings are in the public domain, these quality audits have materially impacted program quality and the surrounding academic and administrative processes.

IDP is the major single source of students to Australian institutions via its agency activities. It commenced in 1969 as a government-funded organization to strengthen teaching and research in Southeast Asia and later the Pacific. IDP was gradually transformed into a "not-for-profit" company wholly owned by Australia's universities with three major business arms: (a) recruitment of international students, (b) English language testing through its shareholding in IELTS, and (c) management of international aid projects. By 2006, IDP had further transformed itself into a "for-profit" company with 50 percent equity held by commercial interests and 50 percent by universities. IDP has also undertaken major industry research particularly in terms of global student circulation over a 12-year period.

Recruitment agents, not only IDP, have been important to the growth of the international student program in Australia, accounting for nearly

60 percent of beginning students in the university sector and higher levels in some other sectors. The two major destination cities, Sydney and Melbourne (and, to a lesser extent, other state capitals), have became international student hubs for local agents with branches in Asia, private English and diploma pathway colleges, and city-based private campuses of universities, some not otherwise present in the city or state. Australian Education International research (2008a, 5) has shown that 35.5 percent of beginning international students come through study-pathway programs, a major difference between Australia and other export-based countries.

By the mid-1990s, most Australian universities had begun to formalize their international strategies and to bring them within the overall university strategic framework. As part of this activity, the non-commercial components of internationalization began to receive greater attention, in particular the development of cooperative linkages and networks, internationalization of the curriculum, and student mobility.

One early aspect of internationalization in universities and the activities of IDP was participation in aid development projects. The emergence of commercial project companies, partly through the operation of the Australia–United States Free Trade Agreement, means that this activity has all but disappeared.

The Australian Universities Directors Forum (AUIDF), an informal grouping of international directors, carries out annual benchmarking of university international activities, primarily of university international marketing and recruiting costs and performance (and, less frequently, student mobility). These benchmarks have become important ways in which university international offices can judge aspects of their performance against industry norms, and have contributed materially to the professionalism of the activity. In its 2007 mobility study (Olsen 2008), using a comparable methodology to studies such as Open Doors, the AUIDF calculated the number of international study experiences (10,718) as a percentage of the graduating cohort of the 37 participating universities. It showed that 5.4 percent of completing undergraduate students had an international experience, and that there were several universities above 10 percent (with one at 18 percent). The study also reported that 60 percent of outgoing experiences had university financial support, 4 percent by the government-based University Mobility in the Asia Pacific (UMAP) scheme and 13 percent by OS-HELP (a government-based loan scheme). The amount of scholarship support provided by universities was approximately US$8 million. Although not comparable to the massive support provided by EU initiatives such as ERASMUS, it shows a clear commitment by universities.

By November 2008, there were some 538,000 international students enrolled in Australian institutions (AEI 2008b). These included 183,000 in universities, 173,000 in private and public vocational education programs, 122,000 in English language programs, and 31,000 in schools. This amounted to export income of some A\$14 billion (US\$10 billion) making education the third export industry in Australia, and accounting for some 6.5 percent of total student circulation globally (OECD 2007). In addition, there were over 60,000 students enrolled in Australian universities in offshore locations, 42 percent in China and India. Within higher education, international students comprise 19 percent of enrollments, the highest portion of any OECD country (Sushi 2008).

The program has not been without its critics, largely around English language issues, the skilled migration program, concerns about the lack of social inclusion of students, sustainability, and teaching and learning issues (Birrell 2006; Marginson 2008; Ross 2009). In addition, a number of universities have decided to move to zero growth of international students, given concerns about the ability to absorb more than 20–25 percent of the university's load from international students.

It has been suggested that international education in Australia has progressed through three phases (Sushi 2008). The first phase (aid funding) progressed from the 1950s to the mid 1980s; the second (high growth) progressed from 1987 to the early years of the new millennium through recruitment of fee-paying students; and the third phase, a more balanced and sustainable approach, has now commenced. This is both a reflection on, and a response to, issues that are also influencing Europe.

Universities have developed significant programs in student and staff mobility, institutional collaboration, and research. International education in Australian universities has become multi-dimensional and has been increasingly encouraged and supported by government attitudes and policies. An appropriate definition of this third phase might be as follows: international education in Australia is centered on trade in educational services, both onshore in Australia and transnationally, with rigorous government intervention in terms of consumer protection and quality assurance.

# Conclusion

The most striking trend over the past 40 years is the increase in the number of globally circulating students, from approximately 250,000 in 1965 to 2,500,000 in 2005. UNESCO (2006, 34) observes a first wave in the period 1975–1985 with an increase of 30 percent (from 800,000 to 1,000,000), a second wave between 1989 and 1994 (with an increase of 34 percent), and a third wave between 1999 and 2004 (with an increase of 4 percent). The most striking recent development is that the traditional destination countries for international students, The United States, United Kingdom, Germany, France, and Australia face increasing competition from countries like China, Singapore, and Malaysia. Countries that send large numbers of students abroad increasingly also become recipients of international students.

Global competition for highly skilled manpower is becoming a strong pull factor in international student circulation. The graying societies of Europe are competing with North America, Australia, and Japan for top talent around the world, all of which need to fill the gaps in their knowledge economies. At the same time, they have to compete with the emerging economies in Asia, Latin America, and Africa, who perhaps need such talents even more.

The cross-border delivery of higher education, with programs, projects and providers moving across borders instead of students, is an important growth market for Australia and the United Kingdom, while continental Europe lags behind. The number of students in offshore activities for both the United Kingdom and Australia are increasing more rapidly than onshore.

Within this context, the internationalization of education in Europe and Australia has, by different paths, time scales, and degrees, moved to a closer set of priorities and actions than formerly. In Europe, trade in educational services is becoming important within its culture of cooperation and mobility (although much more strongly in the United Kingdom than in continental Europe). In Australia, there is a growing balance between trade-dominated activity and cooperation and mobility.

Issues that have historically been part of Australian international education, such as the appointment of agents and developing country marketing plans, are now becoming increasingly important in Europe, while cooperative activities that have been dominant in Europe, for example, the mobility of domestic students to better prepare them for life in a globalized society, have become important in Australia. Our comparison of Australia and Europe illustrates the diverse way higher education responds to the increasingly more international competitive environment. The higher education subsector in Australia and the United Kingdom had by the mid-1980s shifted from aid to trade in their international orientation. In continental Europe this shift has been less radical, taken more time, and occurred via a shift from aid to cooperation and exchange first, before moving toward competition.

# References

Adams, Tony. 1998. "The Operation of Transnational Degree and Diploma Programs: The Australian Case." *Journal of Studies in International Education* 2 (1): 3–22.

———. 2004. "An International Update: The Development of a Long Term Sustainable, Service Business." Sydney: Macquarie University Council Papers.

Altbach, Phillip. 2006. "Globalization and the University: Realities in an Unequal World." In *International Handbook of Higher Education,* ed. P. G. Altbach and J. J .F. Forest. Dordrecht, The Netherlands: Springer.

Australian Department of Immigration and Citizenship. 2008. *Assessment Levels and the Student Visa Program.* Canberra: Department of Immigration and Citizenship. http://www.immi.gov.au.

Australian Education International. 2008a. *Study Pathways of International Students in Australia.* Canberra: Australian Education International. http://aei. gov.au.

———. 2008b. *Monthly Summary of International Student Enrolment Data-Australia, November 2008.* Canberra: Australian Education International. http://aei.gov.au.

Back, Kenneth, and Dorothy Davis. 1995. "Internationalization of Higher Education in Australia." In *Strategies for Internationalization of Higher Education,* ed. H. de Wit. Amsterdam: European Association for International Education.

Barron, Britta. 1993. "The Politics of Academic Mobility in Western Europe." *Higher Education Policy* 6 (3): 50–54.

Birrell, Bob. 2006. *Evaluation of the General Skilled Migration Categories.* Canberra: Commonwealth of Australia.

Chandler, Alice. 1989. *Obligation or Opportunity: Foreign Student Policy in Six Major Receiving Countries.* IIE Research Report No. 18. New York: Institute for International Education.

Cuthbert, Denise, Wendy Smith, and Janice Boey. 2008. "What Do We Really Know About the Outcomes of Australian International Education?" *Journal of Studies in International Education* 12 (3): 255–275.

de Wit, Hans. 2002. Internationalization of Higher Education in the United States of America and Europe: A Historical, Comparative, and Conceptual Analysis. Westport, CT: Greenwood Press.

———. 2005. "Report on Internationalization (E21)." Report for the European Benchmarking Programme on University Management, Brussels.

———. 2008a. "The Internationalization of Higher Education in a Global Context." In *The Dynamics of International Student Circulation in a Global Context,* ed. P. Agarwal, M. E. Said, M. Sehoole, M. Sirozi, and H. de Wit. Rotterdam, The Netherlands: Sense Publishers.

———. 2008b. "International Student Circulation in the Context of the Bologna Process and the Lisbon Strategy." In *The Dynamics of International Student*

*Circulation in a Global Context,* ed. P. Agarwal, M. E. Said, M. Sehoole, M. Sirozi, and H. de Wit. Rotterdam, The Netherlands: Sense Publishers.

Frolich, Nicoline, and Amelia Veiga. 2005. "Competition, Cooperation, Consequences and Choices in Selected European Countries." In *Internationalization in Higher Education: European Responses to the Global Perspective,* ed. B. Khem and H. de Wit. Amsterdam: European Association for International Education and the European Higher Education Society (EAIR).

Knight, Jane. 2008. *Higher Education in Turmoil. The Changing World of Internationalization.* Rotterdam, The Netherlands: Sense Publishers.

Marginson, Simon. 2008. "Sustainability and Risks of Internationalization." Paper presented at the Australian Financial Review Higher Education Conference, Sydney.

Naidoo, Rajani, and Ian Jamieson. 2005. "Knowledge in the Marketplace: The Global Commodification of Teaching and Learning in Higher Education." In *Internationalizing Higher Education, Critical Explorations of Pedagogy and Policy,* ed. P. Ninnes and M. Hellsten. Hong Kong: Comparative Education Research Centre, the University of Hong Kong/Springer.

Organisation for International Co-operation and Development. 2007. "Education at a Glance." In *OECD Indicators.* Paris: OECD.

Olsen, Alan. 2008. "Outgoing International Mobility of Australian University Students 2007." Australian Universities International Directors Forum, Canberra, Australia.

Reinalda, Bob, and Ewa Kulesza. 2005. *The Bologna Process—Harmonizing Europe's Higher Education.* Opladen, Germany, and Bloomfield Hills, MI: Barbara Budrich Publishers.

Ross, John. 2009. "Australia Now Playing It Safe on Safety: Nyland." *Campus Review, 2 February.* Sydney: Fairfax Press.

Scott, Peter. 2005. "The Global Dimension: Internationalising Higher Education." In *Internationalization in Higher Education: European Responses to the Global Perspective,* ed. B. Khem and H. de Wit. Amsterdam: European Association for International Education and the European Higher Education Society (EAIR).

Sushi, Das. 2008. "Foreign Students at Capacity Levels." *The Age,* July 26. Melbourne, Australia.

Teichler, Ulrich. 2004. "The Changing Debate on Internationalization of Higher Education." *Higher Education,* 48: 5–26.

United Nations Education, Scientific, and Cultural Organization. 2006. *Global Education Digest 2006.* Paris: United Nations Education and Scientific Organization.

van der Wende, Marijk C. 2001. "Internationalization Policies: About New Trends and Contrasting Paradigms." *Higher Education Policy* 14 (3): 249–259.

van Vught, Frans, Marijk van der Wende, and Don F. Westerheijden. 2002. "Globalization and Internationalization. Policy Agendas Compared." In *Higher Education in a Globalizing World: International Trends and Mutual Observations. A Festschrift in Honor of Ulrich Teichler,* ed. J. Enders and O. Fulton. Dordrecht, The Netherlands: Kluwer Academic.

# Chapter 17

# Student Mobility and Emerging Hubs in Global Higher Education

*Robin Shields and Rebecca M. Edwards*

Bangalore, India, has become an icon of globalization and the rise of the knowledge economy. Virtually every major technology company has opened an office there, seeking to capitalize on the large supply of highly skilled labor that remains relatively inexpensive by global standards. In a juxtaposition of old and new, foreign executives stay in five star hotels while oxen pull vegetable carts in the streets outside. Among the gleaming information technology (IT) office parks, another industry is being born: young foreigners are flocking to Bangalore—not on business, but rather to pursue their own education. The same universities that gave birth to India's IT revolution have become recognized centers of scientific and engineering education. Students from Iran, Saudi Arabia, and the United Arab Emirates come to the area in growing numbers to receive an education that many consider to rival the best universities in the world at a fraction of the cost.

In this chapter, we examine how emerging global hubs in the higher education "industry" are transforming patterns of international student mobility and competition by attracting a growing number of international students. We consider the relationship between global competition and international student mobility as one that is multidirectional and recursive: recent growth in student mobility has played a role in increasing competition, which, in turn, has reshaped student mobility patterns. This process of change is deeply connected to—and implicated with—globalization, the development of new economies, and the flow of human resources across borders, creating a dynamic higher education environment that has yet to be defined and understood.

We begin by describing the changing patterns of global student mobility, focusing on new destinations that are attracting an increasing number of international students. We then describe measures that well-established universities are taking to protect their global preeminence and contrast these to the emerging competitive hubs. Our analysis suggests that the degree to which most universities are competitive in recruiting international students will increasingly be shaped by the scope and character of their global network rather than any intrinsic characteristics of their individual courses, culture, or campuses.

# The Rise of International Student Mobility

The international flow of university students is not a new phenomenon: Baiba Rivza and Ulrich Teichler (2007) estimate that 10 percent of students at European medieval universities were international. Since the mid-twentieth century, the flow of students across international borders began to grow at an unprecedented rate: foreign student enrollment worldwide has grown from 107,589 in 1950 to 2.5 million in 2008 (Barnett and Wu 1995; Gürüz 2008).

Two critical factors have contributed to this rapid growth. First, the second half of the twentieth century saw an increase in higher education enrollment around the world. Global enrollment in higher education increased from six million in 1950 to nearly 132 million in 2004 (Gürüz 2008). This rise in enrollment was closely linked to the advent of the "knowledge economy," which meant that a greater share of economic activity began to center on the exchange of information (Drucker 1968). The shift to knowledge-based production required a workforce with scientific and technical expertise, increasing the demand for higher education on both the individual and international levels.

Second, student exchange programs became integral to the international development sector and the foreign policy agendas of many countries. Liping Bu (1999) points out the connection between student exchange and international development with the observation that the "thirty-three universities with the highest foreign student enrollment (42 percent of total foreign student population) were also most heavily involved in...university contracts for foreign aid" (405). Similarly, during the Cold War both the United States and the Soviet Bloc utilized student exchanges to maintain and garner international support and project images of their power and prosperity (Barnett and Wu 1995).

Almost without exception, the flow of international students followed a pattern of students moving from the global "periphery"—"third world" or "developing" countries—to the "core"—industrialized, "first world," "developed" countries (Altbach 1989).[1] Among the core countries, five Major English Speaking Destination Countries (MESDCs)—the United States, United Kingdom, Australia, Canada, and New Zealand—alone accounted for 47 percent of international students' enrollment in 2003 (Böhm et al. 2004). This pattern of student mobility from periphery to core countries has been widely accepted as a mutually beneficial, or symbiotic, relationship: destination countries gain by advancing their foreign policy objectives, accessing skilled labor, and (increasingly) funding their own higher education institutions (HEIs), while sending countries access new skills and knowledge networks (Altbach 1989).[2] Critics might argue that the effect of "brain drain" essentially negates the latter benefits because a large number of foreign students eventually take up permanent residence in their host countries. However most foreign students, who settle abroad retain strong ties to their home countries and, in some cases, have been instrumental in building the capacity of their higher education systems at home (*The Economist* 2005).

## Changes in Student Mobility: Competition and Globalization

While international student mobility continues to rise, the past decades have seen significant changes in higher education brought about by increased competition and globalization. Together, these forces have played a significant role influencing patterns of student mobility.

Higher education systems in most major destination countries have become increasingly open to free market mechanisms. This change coincided with the privatization and economic liberalization of social services that occurred simultaneously in many countries during the 1980s, including a decrease in state funding of higher education (Le Grand and Bartlett 1993; Dolenec 2006). In the higher education subsector, the move toward economic liberalization was justified by the arguments that (a) market forces would improve efficiency and lower the costs of higher education, and (b) higher education represents a private interest and a personal investment as much as a public good (Marginson 2004).

In the United States and United Kingdom, policies were enacted that created the "pervasive introduction of trade concepts, language, and policy" (Collins 2007, 285) in higher education. The 1980 *Bayh Dole Act* paved the way for universities in the United States to generate increased shares of revenue by licensing intellectual property developed through publicly funded research. Five years later, the United Kingdom instituted its own series of higher education reforms based on the recommendations of the Jarratt Commission, which advocated "managerialism," performance monitoring, and market-based competition mechanisms (Land 2004).

Similar provisions followed on the international level with the 1995 General Agreement on Trade in Services (GATS) of the World Trade Organization. Essentially, GATS classified education as a service sector of the global economy and stipulated that it is open to the same free trade regulations that govern other service exports. The stipulations of GATS have been fiercely contested, with many claiming that it lacks transparency, commodifies a social good without concern for equity, and may destabilize the higher education systems of developing countries. Several higher education associations and interest groups have even issued dissenting statements and declarations, including the American Council on Education, the European University Association, the African Association of Universities, and the United Nations Educational, Scientific, and Cultural Organization (UNESCO) (Collins 2007).

The privatization of higher education has also brought about an increased interest on the part of HEIs in attracting foreign students as a result of revised fee structures. Many countries have adopted a two-tier fee structure, in which residents of the country pay lower, government-subsidized tuition fees while international students pay fees that effectively represent the "full cost" of their education (Cummings 1984; Gürüz 2004). As a result, competition for foreign students has become fierce, with many universities opening specialized recruitment offices in major sending countries.

Simultaneous with the economic liberalization and privatization of higher education, globalization has broken down the division of "producers" and "consumers" in the knowledge economy. Global knowledge networks have moved away from a simple core-periphery structure, in which countries whose economies relied heavily on services and manufacturing (e.g., China, India, and Japan) have become knowledge producers in their own right. As a result, "third world" institutions such as the Indian Institutes of Technology—which have alumni sitting on the boards of top U.S. technology companies and teaching in Ivy League universities—have gained recognition as preeminent centers in so-called STEM fields (i.e., science, technology, engineering, and mathematics).

As a result of competition and globalization, patterns of student mobility are beginning to change. While MESDCs retain a strong share of international student enrollment, recent years have also witnessed the appearance of new destinations for international students. Some of these locations, such as Japan, India, and China, have long-running higher education systems that only recently gained global popularity as destinations for foreign students. Others (e.g., the Gulf States) are rapidly investing in internationally competitive higher education systems as part of their larger national development agendas. Many have increased English-medium offerings in part to further their competitiveness.

Data on international student mobility is difficult to compare and analyze due to varying definitions of higher education enrollment, incomplete data sources, lack of extensive historic data, and the rapidly changing nature of the field. The UNESCO Institute for Statistics' data center and its associated *Global Education Digests* include extensive data on student mobility trends, including outbound and inbound enrollment figures, gender ratios, and major destinations for students from each country. Similarly, the Institute of International Education's *Atlas of International Student Mobility* compiles data from a number of sources, including ministries of education, private consulting firms, and UNESCO's data.

Table 17.1 presents data compiled from both sources to illustrate varying average annual growth rates in international student enrollment in select countries between 2002 and 2007.

In most countries, international enrollment has grown, in many cases at a relatively rapid rate. However, the growth in emerging destinations (e.g., South Korea, India, and China) clearly outpaces that of more established destination countries (e.g., the United Kingdom, France, and Germany).

This rise of new destinations for international students is unlikely to usurp the global dominance of traditionally preeminent universities overnight. However, it does indicate that competition will increase for the revenues associated with foreign student enrollment. Alison Wride, a professor at the University of Exeter's School of Business and Economics, recently predicted a "downturn in 10–15 years in the international student market [in the United Kingdom] as China and Pakistan further develop their own [academic] infrastructures," noting that "if we are to continue to compete in a global market, we need to ensure that what we offer is world-class education" (Engage 2008, 7). Indeed, "world-class" appears to increasingly entail a global presence and reach, as demonstrated by the bold and entrepreneurial new measures that universities in MESDCs are taking to protect their global preeminence.

**Table 17.1** Comparison of Average Annualized Growth Rates in International Students

| | 2002 | 2003 | 2004 | 2005 | 2006 | 2007 | Average Annualized Growth Rates (%) |
|---|---|---|---|---|---|---|---|
| Australia[a] | 179,619 | 188,160 | 166,954 | 207,264 | — | 211,526 | 6.2 |
| Canada[a] | 52,596 | — | 70,023 | 75,546 | — | — | 7.9 |
| China[b] | | 77,715 | 110,844 | 141,087 | | 162,695 | 30.1 |
| France[a] | 165,437 | 221,567 | 237,587 | 236,518 | 247,510 | 246,612 | 15.0 |
| Germany[b] | — | 227,026 | 246,300 | 246,334 | 248,357 | 246,369 | 2.8 |
| India[b] | | | 7,589 | 13,267 | | 18,594 | 47.4 |
| Japan[a] | 74,892 | 86,505 | 117,903 | 125,917 | 130,124 | 125,877 | 14.7 |
| South Korea[a] | 4,956 | 4,956 | 7,843 | 15,497 | 22,260 | 31,943 | 60.7 |
| New Zealand[a] | 17,732 | 26,359 | 41,422 | 40,774 | — | 33,047 | 47.4 |
| United Kingdom[a] | 227,273 | 255,233 | 300,056 | 318,399 | 330,078 | 351,470 | 9.2 |
| United States[a] | 582,996 | 586,316 | 572,509 | 590,158 | 584,719 | 595,874 | 0.8 |

*Notes*: a UNESCO Institute for Statistics 2009.
b IIE 2009.

Data are taken from multiple sources, thus direct comparison of gross enrollment figures is not advisable, though comparison of growth rates is possible. Average annual growth rates were computed by authors as an average of year-on-year growth figures with gaps in data interpolated.

# Preeminent Institutions in Jeopardy:
# Emerging Efforts to Compete

As a result of growing competition for international student enrollment, institutions in established destination countries have begun to experiment with the delivery models they use to keep up with the increasing number of competing options available to international students. While several of these models are not necessarily new, the extent to which they have grown and diversified in the last decade speaks to the emphasis these institutions have placed on maintaining a share of the international student market. Stuart Cunningham et al. (1998) characterize these developments as "borderless" in nature. Among the new delivery models that are emerging, those that may have a particularly significant impact on student mobility patterns fall under the GATS framework Mode 3 of "commercial presence," in which providers use a variety of means to establish physical presence in foreign countries. Tony Adams (1998) provides a useful typology for these forms, including twinning, franchising, offshore or branch campuses, moderated programs, joint degree awards, and distance learning. All of these approaches include elements of "offshoring," or creating a physical presence in outside countries, and many include some degree of cross-campus partnerships.

Offshoring approaches are on the rise, both in terms of delivery efforts and student enrollment. Australian HEIs provide a particularly notable example in that international students have played a critical role in these institutions' growth. In the ten years between 1996 and 2005, the number of students enrolled in offshore programs run by Australian universities jumped from 10,483 to 63,906, an increase of over 600 percent. This outpaces even the phenomenal growth in foreign students in Australian universities as a whole (including offshore programs), which climbed from 8 to 25 percent of the overall student body (Niland 2008, 79). Similarly, enrollment in British offshore programs increased from about 120,000 to 190,000 students in just the three years between 2000 and 2003 (Gürüz 2008, 108). The importance of satellite campuses as a new offshore delivery model can be seen in the growth of educational hubs such as Qatar's Education City and Dubai's Knowledge University, large campuses created by the respective governments for the sole purpose of housing satellite campuses of foreign universities.

Other forms of offshore delivery have been less successful: online distance education, once heralded as the future of the higher education subsector, has failed to make a serious impact in student mobility to date. Institutions such as New York University and Michigan State University, which launched online distance education programs (often branded "virtual universities" or "online campuses") in the late 1990s, closed their operations several years later due to lack of demand and financial shortfalls (Gürüz 2008).

The rise in demand for offshore programs offered by established institutions will likely continue to impact student mobility flows and global competition in general. This may entail a shift from "periphery-core" patterns to "periphery-periphery," as many students seeking degrees from reputed international institutions will be able to get them by traveling within the local region (and perhaps without even leaving home). As these patterns shift, incoming student numbers may become less important as institutions in established destination countries focus on revenue generation through offshore enrollment. However, these new delivery models often involve some overseas study as part of the degree gaining process, with students leaving their home countries for only a semester or a year, rather than taking up long-term residence abroad.

By establishing branded campuses and degree programs abroad, core institutions may maintain a share of international student enrollment and remain competitive in the global higher education market. However, this may be something of a subdued victory: while globally renowned institutions may maintain the reputation and preeminence of their "brand," they may lose some of the financial incentives associated with international students. In particular, the practice in some countries of effectively subsidizing home students through international enrollment will be difficult to maintain as offshore campuses emerge as recognized institutions in their own right and have to contend with other competitive pressures on the regional level.

## Emerging Hubs and Global Networks: The New Competition

While universities in established destination countries are looking to new delivery models and international partnerships to deal with increased global competition, those in emerging hubs are developing responses oriented toward global competition and the international student market. These strategies seek to situate their universities at the crossroads of global knowledge and economic networks, offering students an opportunity to develop connections, skills, and expertise that effectively bridge the core and periphery. In this section, we will explore two different examples of countries that are investing heavily in a new class of institutions that aim to be elite and global in their appeal, relevance, and connections: Saudi Arabia and Singapore.

## Saudi Arabia's Aim to Be Elite: The King Abdullah University of Science and Technology

Saudi Arabia's King Abdullah University of Science and Technology (KAUST) offers one example of how universities in emerging destinations are taking entrepreneurial steps to create local hubs in global knowledge networks. As a new institution, KAUST represents a bold attempt to create a globally competitive university overnight by establishing a top-notch, cutting-edge research university offering graduate-level (i.e., PhD and MS) degrees in a number of STEM fields, with all courses taught in English.

Integration into global knowledge networks forms a cornerstone in KAUST's development strategy, an objective that it plans to pursue through a combination of research partnerships, faculty recruitment, and international student enrollment. Utilizing its multibillion U.S. dollar endowment from Saudi Arabia's King Abdullah, KAUST has the financial resources to offer new faculty seed funding for global research partnerships with other leading universities. In a radio interview, Provost Fawwaz Ulaby described this as an effort to build "bridges with academic institutions around the world" (Abramson 2008). To complement these academic networks, the university's international advisory council, with representatives from a number of top international companies, is well-positioned to forge strategic links with private sector companies.

KAUST's emphasis on its international network suggests that it hopes to construct itself as both a global university and a Saudi institution. Furthermore, it is pursuing this global status not simply through joint delivery or exchange programs, but also by aiming to become a global knowledge hub in its own right. The strength the university is building will ultimately lie in its success in creating a strategic position within a worldwide network of universities, private sector organizations, and government entities.

## Singapore's Strategy of Diversification

In the last decade, Singapore has also begun to develop and globalize its higher education system through a rapid, extensive strategy that stresses diversification and entrepreneurialism, and which includes elements to attract foreign students. The intention behind this approach has been to create a regional hub of high quality talent, research, and economic development with a competitive edge in Asia both educationally and economically (Tan 2004).

Singapore's strategy of diversification is best illustrated by reviewing some recent developments. These advances include the introduction of a new, private university in 2000 (Singapore Management University) based on the American model of education, and the 2006 corporatization of Singapore's national universities. Singapore has also opened itself up to foreign investments in higher education from institutions such as France's Institut Européen d'Administration des Affaires (INSEAD), or European Institute for Business Administration) to help make its system more competitive. Yet another approach to diversification involves sending Singaporean students abroad, in this case not to foreign institutions, but rather to "offshore" Singaporean satellite campuses in technological growth centers such as Silicon Valley, Bangalore, and Shanghai. This strategy was developed to nurture the development of a Singaporean workforce familiar with competing economies and with networks within these economies that could support entrepreneurialism and even their own startup businesses. Finally, Singapore has sought to position itself more effectively in the global marketplace by developing the Campus for Research Excellence and Technological Enterprise (CREATE), a Singapore-based research and development hub with partners from around the world, including the Massachusetts Institute of Technology.

The examples of Saudi Arabia and Singapore evince a common strategic approach: in both cases, the universities' primary *raison d'être* remains to provide domestic students with higher education, yet they have both recognized that integration into global mobility streams offers a competitive advantage that cannot be ignored. The success of these universities depends on their integration into global networks, both so that their students can become professionally successful subsequent to graduation, and so that the institutions themselves can gain access to, and take part in, the broader global project of knowledge creation. In fact, in some cases, the logic for this global orientation becomes almost circular: a university's status and recognition as a high quality global institution is largely determined by its ability to attract international students, which in turn furthers its global reputation.

## Analysis: Universities as Networks

In her work on global educational policy borrowing and lending, Gita Steiner-Khamsi (2004) suggests that studies of international educational policy reform should give heed to theories of network analysis. Rooted in systems theory and complexity theory, social network analysis avoids the description of deterministic cause and effect relationships and instead seeks to explain how the choices of individual actors within a system lead to emergent, macro-level characteristics of the system of a whole, which may be quite different from the characteristics of any of its constituent actors.

This perspective seems particularly *apropos* in the field of global higher education where hierarchical distinctions between established universities and those in the periphery that seemed so deeply ingrained and pervasive only a few decades ago are quickly becoming irrelevant. As Steiner-Khamsi (2004) puts it, "distinctions…between local and global, internal and external…perpetuate erroneous dichotomies and need to be dismissed altogether" (214). Thus, complex networks of knowledge, information, and resources that encompass both the core and periphery are becoming more powerful. Rather than simply through institutional reputation, the competitiveness of HEIs will increasingly be determined through their integration into the network, not only by key partnerships, but also through the number of "weak ties" that they hold. Weak ties, which are casual or informal links, play a vital role in the propagation of changes to policy and practice throughout the network. According to sociologist Mark Granovetter (1983), "social systems lacking in weak ties will be fragmented and incoherent. New ideas will spread slowly, scientific endeavors will be handicapped, and subgroups separated by race, ethnicity, geography, or other characteristics will have difficulty reaching a *modus vivendi*" (202). Steiner-Khamsi (2004), too, notes that "global players" in educational policy reform are able to "maintain 'weak ties' to different clusters of stakeholders that that are spread throughout the globe" (215).

Whether the global university system becomes entirely unhinged from its "core" to "periphery" axis remains to be seen, as many factors influence patterns of student mobility. Even with a radical reorientation of knowledge networks, the reputations of the Ivy Leagues and Oxbridge will not disappear overnight. Higher education decisions are part of a student's overall life strategy, and factors external to the educational experience (e.g., the potential for work experience and permanent residency after study) will continue to impact student mobility patterns. However, the next decades may well witness a transition to a global higher education system that is far less hierarchical, less dichotomous in nature, and more oriented toward global and institutional networks.

## Conclusion

In this chapter, we have demonstrated how changes to international student mobility are impacting competition in the higher education subsector.

A number of new hubs are attracting increasing numbers of international students, who have become an important source of revenue for universities in many countries. As a result, many of the more established universities have been forced to engage with new delivery models in order to compete with these emerging hubs of higher education. We argue that the outcomes of these new delivery models may in some cases be uncertain, and that universities' global networks, specifically their ties to other universities, private sector organizations, and governments, will take on an increasingly important role in determining their success.

How this increased competition affects "on the ground" realities of higher education remains to be seen. While we have focused our analysis on the international level, competition related to student mobility will also have ramifications within countries. Just as universities and national governments have pursued distinct strategies for international student recruitment, they will have to contend with new competitive pressures in their own unique ways. Moreover, many will face challenges associated with the risks of these new endeavors.

Elite institutions in "core" countries, whose names carry a connotation of superiority with global recognition, likely have a reputation that will endure any amount of international competition. Even with the emergence of new hubs, they will likely be able to attract large numbers of international students on whatever terms they choose. Other universities in MESDCs will be in a far more precarious situation. Those which focus primarily on teaching (rather than research) and those which currently or in the future expect to rely on foreign student fees for the financial stability of the university may have to transform their approach to compete in the international market; some may even have to redefine their very institutional character.

Similarly, on the local level, competition in the higher education subsector will have varying implications for students from different socioeconomic backgrounds. The scope of competition in higher education has grown beyond what even the staunchest supporters of "marketization" could have foreseen just 20 years ago, and equity and social justice may not get due attention in this global "free for all." On one hand, increased competition may offer more students the opportunity to receive a "globally competitive" education at a cost that they can afford. On the other, marginalized groups who struggle for access to higher education may continue to face many of the same difficulties, risking a vicious cycle of social exclusion. The true measure of whether competition truly "works" may ultimately be in whether it can expand opportunities for the masses while maintaining the prospects of social mobility and innovation that have driven the phenomenal growth of universities around the world.

# Notes

1. In our analysis, we use the terms "core" and "periphery" to categorize major destination and source countries, respectively. While doing so, we acknowledge these terms and their synonyms (e.g., "North" and "South") are deeply problematic. Indeed, one of our central arguments is that such dichotomous distinctions are decreasingly relevant in the field of higher education. While these terms are contested, we intend the "core" to include the five Major English Speaking Destination Countries (MESDCs) and, to a lesser extent, Western European countries such as France and Germany. In contrast, the "periphery" refers to most other countries, especially those that send large number of students abroad every year (e.g., India, China, Pakistan, and much of Southeast Asia).
2. The authors recognize that the role of international students in funding institutions varies among countries and between levels (e.g., undergraduate, master's, and doctoral).

# References

Abramson, Larry. 2008. "Premier Research University Rises in Saudi Desert." *Morning Edition*, National Public Radio, August 4.

Adams, Tony. 1998. "The Operation of Transnational Degree and Diploma Programs: The Australian Case." *Journal of Studies in International Education* 2 (3): 3–22.

Altbach, Philip G. 1989. "The New Internationalism: Foreign Students and Scholars." *Studies in Higher Education* 14 (2): 125–136.

Barnett, George A., and Reggie Y. Wu. 1995. "The International Student Exchange Network: 1970 & 1989." *Higher Education* 30 (4): 353–368.

Böhm, Anthony, et al. 2004. *Vision 2020, Forecasting International Student Mobility: A UK Perspective.* London: British Council.

Bu, Liping. 1999. "Educational Exchange and Cultural Diplomacy in the Cold War." *Journal of American Studies* 33 (3): 393–415.

Collins, Christopher S. 2007. "A General Agreement on Higher Education: GATS, Globalization, and Imperialism." *Research in Comparative and International Education* 2 (4): 283–296.

Cummings, William K. 1984. "Going Overseas for Higher Education: The Asian Experience." *Comparative Education Review* 28 (2): 241–257.

Cunningham, Stuart, Suellen Tapsall, Yani Ryan, Lawrence Stedman, Kerry Bagdon, and Terry Flew. 1998. *New Media and Borderless Education.* Canberra: Department of Employment, Education, Training and Youth Affairs.

Dolenec, Danijela. 2006. "Marketization in Higher Education Policy: An Analysis of Higher Education Funding Policy Reforms in Western Europe between

1980 and 2000." *Revija za socijalnu politiku* [Review of Political Sociology] 13 (1): 15–35.

Drucker, Peter. 1968. *The Age of Discontinuity: Guidelines to Our Changing Society.* New York: Harper and Row.

*The Economist.* 2005. "The Brains Business." *The Economist* U.S. Edition, September 10: 3–4.

*Engage.* 2008. "A Career in Higher Education: Inspiration, Opportunity, Strategy, Luck?" *Engage* 15 (Autumn): 6–7.

Granovetter, Mark. 1983. "The Strength of Weak Ties: A Network Theory Revisited." *Sociological Theory* 1: 201–233.

Gürüz, Kemal. 2008. *Higher Education and International Student Mobility in the Global Knowledge Economy.* Albany, NY: State University of New York Press.

Institute of International Education. 2009. *Atlas of International Student Mobility.* New York: Institute for International Education. http://www.atlas. iienetwork.org.

Land, Ray. 2004. *Educational Development: Discourse, Identity, and Practice.* Maidenhead: Society for Research into Higher Education and Open University Press.

Le Grand, Julian, and Will Bartlett. 1993. *Quasi Markets and Social Policy.* London: Macmillan Press.

Marginson, Simon. 2004. "Competition and Markets in Higher Education: A 'Glonacal' Analysis." *Policy Futures in Education* 2 (2): 175–244.

Niland, John. 2008. "The Engagement of Australian Universities with Globalization." In *The Globalization of Higher Education*, Glion Colloquium Series no. 5, ed. L. E. Weber and J. J. Duderstadt. London: Economica.

Rivza, Baiba, and Ulrich Teichler. 2007. "The Changing Role of Student Mobility." *Higher Education Policy* 20 (4): 457–475.

Steiner-Khamsi, Gita. 2004. "Blazing a Trail for Policy Theory and Practice." In *The Global Politics of Educational Borrowing and Lending*, ed. G. Steiner-Khamsi. New York: Teachers College Press.

Tan, Jason. 2004. "Singapore: Small Nation, Big Plans." In *Asian Universities: Historical Perspectives and Contemporary Challenges*, ed. P. G. Altbach and T. Umakoshi. Baltimore: Johns Hopkins University Press.

UNESCO. 2009. *Statistics.* Paris: UNESCO. http://stats.uis.unesco.org.

# Notes on Contributors

**Sylvia S. Bagley** is the Fritz Burns Endowed Professor in Education at Mount Saint Mary's College in Los Angeles, where she is Director of the Master's Program in Instructional Leadership. She was formerly Co-Administrative Director of the Center for International and Development Education at UCLA, where she coordinated country-specific research for CIDE's Higher Education Database and worked with a local non-profit organization (Relief International) to provide international teacher exchange programs. Previously, as Academic Integration Specialist for the Education Abroad Program at UCLA, she worked with subject-area counselors to help align UCLA's course requirements with courses taken abroad. Her current research interests include authentic assessment methods, K-12 teacher development, lifelong learning habits, student identity, and a comparative approach to post-secondary alternative schooling.

**Laura M. Portnoi** is Assistant Professor in the Advanced Studies in Education and Counseling Department of the College of Education at California State University, Long Beach (CSULB). She serves as the program director for the Social and Cultural Analysis of Education master's program and leads the Graduate Culture Initiative for the College of Education. Prior to joining the faculty at CSULB, she worked in the field of international education exchange for eight years, including a term as Resident Director in South Africa. She holds a bachelor's degree from the University of Michigan, master's degrees from Indiana University and the University of the Western Cape (South Africa), as well as a PhD in Comparative Education from the University of California, Los Angeles. Her research experience and interests pertain to the intersection of higher education and comparative international education. Specific areas of focus include policy borrowing and implementation processes of employment equity policies at higher education institutions in South Africa, the career decision-making processes of graduate students who are poised to pursue employment in academia, and the academic experiences of first-generation graduate students.

**Val D. Rust** is Professor of Education at the University of California, Los Angeles (UCLA) and is currently the Faculty Chair of the Graduate School of Education and Information Studies. He received his PhD from the University of Michigan in Education Studies and currently serves as the Director of the International Education Office at UCLA, which houses the Education Abroad Program, the Travel Study Program, non-University of California study abroad providers and other student exchange programs. He is also Associate Director of the Center for International and Development Education in the Department of Education, which deals extensively with higher educational mapping around the world, international educational leadership, and teacher training.

**Tony Adams** is Director and Principal Consultant of Tony Adams and Associates, Consultants in International Education. He is Co-editor of the *Journal of Studies in International Education* and has published several articles about international education. He is the immediate Past President and Founding President of the International Education Association of Australia and recipient of the Charles Klazek Award from the Association of International Education Administrators (AIEA) and IDP Education Australia for his contribution to international education. He has worked in senior positions in international education at RMIT University in Melbourne and Macquarie University in Sydney, and has an ongoing association with Universita Cattolica Sacre Cuore in Milan.

**Thuwayba Al-Barwani** is Dean of the College of Education at Sultan Qaboos University in Oman and a member of the Oman State Council (parliament). She previously served as Deputy Minister for Social Development and was a member of the Oman Council for Higher Education. Her areas of specialization include education policy and literacy.

**Hana Ameen** is Advisor for Academic Affairs to the Minister of Higher Education in Oman. Her area of specialization is statistics, and she has held a number of faculty positions in departments of statistics in Oman and other countries in the Gulf Region.

**David W. Chapman** is the Rodney Wallace Professor of Education at the University of Minnesota, where he teaches graduate courses in comparative and international development in the Department of Educational Policy and Administration. His specialization is in international development assistance. He has worked in over 45 countries for the World Bank, USAID, UNICEF, Asian Development Bank, InterAmerican Development Bank, UNESCO, and similar organizations. He has authored or edited 7 books and over 125 journal articles and book chapters, many of them on issues related to the development of education systems in international settings.

**Colleen Coppla** is the Executive Director of Campaigns at Fairleigh Dickinson University and a PhD candidate at Seton Hall University in the Higher Education Management, Leadership, and Policy Program. Her research interests include academic capitalism, academic entrepreneurship, and international comparative education.

**Hans de Wit** is Professor, Internationalization of Education at the Amsterdam University of Applied Sciences. He is Co-editor of the *Journal of Studies in International Education*, and a private consultant in international higher education. He was a New Century Scholar of the Fulbright Program in the 2005–2006 Higher Education in the 21st Century Program, and has published widely about international education. He has been Vice-President for International Affairs of the University of Amsterdam. He is consulting for institutions of higher education, the World Bank, UNESCO, IMHE/OECD, and the European Commission, among others. He is a Founding Member and a Past President of the European Association for International Education (EAIE). He has received awards for this work from the University of Amsterdam (2006), AIEA (2006), CIEE (2006 and 2004), NAFSA (2002), and EAIE (1999 and 2008).

**Rebecca M. Edwards** is a PhD candidate in Comparative and International Education at the University of California, Los Angeles. She also holds a master's degree in International Education Development from Teachers College, Columbia University. Her work experience includes educational program management in the U.S. nonprofit sector, and her primary areas of research include education policy, education decentralization, and politics in education, with a regional specialty in Nepal.

**Héctor R. Gertel** is Full Professor of Economics and Education of the Graduate School of Economics at the *Universidad Nacional de Córdoba*, Argentina. He is also Director of the Center for Design, Implementation, and Evaluation of Social Policies. He served for seven years at the Ministry of Education of Argentina as Co-director of the Higher Education Reform Project, which aimed to improve the quality of higher education. He is an editor of several books on the economics of education and has conducted research and evaluation of higher education investment programs in Latin America.

**Li Guo**, a native of China, is pursuing a PhD degree in Higher Education at Seton Hall University. Prior to her current studies, she spent several years in Germany working and studying in the field of vocational education. Her research interest focuses on international comparative education, specifically regarding China as it relates to the world.

**Lynn Ilon** is Associate Professor at Seoul National University's College of Education. She teaches in the Global Perspective in Lifelong Education program. She is an educational economist specializing in educational policy and planning issues arising from the globalization of the world's economies. Ilon specializes in issues around a global knowledge economy—especially as it applies to the value of learning among and between culturally different groups. Ilon has lived in several regions of the world, continues to build her intellectual work through field work globally, and publishes regularly in the areas of global economy, education, and culture.

**Baktybek Ismailov**, Doctor of Technical Science, is the Head of State Inspection at the Ministry of Education of the Kyrgyz Republic. He is also the Vice-President of the Eurasian Quality Assurance Network and Vice-President of the Congress of Educational Employees "Quality Education." He was a Fulbright Scholar of the Bureau of Educational and Cultural Affairs of the U.S. Department of State at Seton Hall University in the 2008–2009 fellowship year. His field of interest is accreditation of higher education institutions.

**Alejandro D. Jacobo** is a Professor of Economics at the Pontificia Universidad Católica Argentina and Universidad Nacional de Córdoba, where he is currently a member of its center for economic research. His areas of interest include the economics of integration and the evaluation of public policies, with an emphasis on higher education quality management. He has received several awards and honors during his professional career.

**Daniel Kirk** is an Assistant Professor at the American University of Sharjah in the United Arab Emirates. His teaching is focused on educational issues in the Middle East, along with the rise of overseas branch campuses, particularly in the Arabian Gulf. His research is comparative in approach, exploring the globalizing forces that are influencing and informing educational policy in the Middle East region. Kirk is the President and founder of the Gulf Comparative Education Society, which aims to bring together those interested in the educational landscape of the region. Kirk taught high school English and Literature for ten years in the United Kingdom, Qatar, Bermuda, and Dubai, before receiving his PhD in Language and Literacy Education from the University of Georgia.

**Jane Knight**, University of Toronto, focuses her research and professional interests on the international dimension of higher education at the institutional, system, national, and international levels. Her work in over 60 countries with UN Agencies, universities, foundations, and professional organizations helps to bring a comparative, development, and international perspective to her research, teaching, and policy work. She is the author of more than 80 articles/chapters/reports on internationalization concepts and strategies, quality assurance, institutional management, mobility, cross-border education, trade, and capacity building. Her latest books (published in 2008) include *Higher Education in Turmoil: The Changing World of Internationalization* (author), *Financing Access and Equity in Higher Education* (editor), and *Higher Education in Africa: The International Dimension* (co-editor). She is an Adjunct Professor at the Ontario Institute for Studies in Education, University of Toronto, a Fulbright New Century Scholar for 2007–2008, and sits on the advisory boards of several international organizations and journals.

**Moeketsi Letseka** is Senior Lecturer in the Department of Educational Studies, College of Human Sciences, University of South Africa (UNISA). His areas of research interest include higher education policy and planning, philosophy of education, educational research, and information communication technology and education.

**Simon Marginson** is Professor of Higher Education in the Centre for the Study of Higher Education at the University of Melbourne, Australia. His work is focused on higher education policy, comparative and international education, and globalization. Current books include *Creativity in the Global Knowledge Economy* (Michael Peters, Simon Marginson, and Peter Murphy, published by Peter Lang, February 2009); *Global Creation: Space, Mobility and Synchrony in the Age of the Knowledge Economy* (Simon Marginson, Peter Murphy, Michael Peters, published by Peter Lang, late 2009); *International Student Security* (Simon Marginson, Chris Nyland, Erlenawati Sawir, and Helen Forbes-Mewett, published by Cambridge University Press, 2010).

**Kathryn Mohrman** is a Professor in the School of Public Affairs at Arizona State University, and became the first director of the University Design Consortium in February 2008, working closely with Sichuan University in China as co-sponsors of the Consortium. Before this position, she was Executive Director of the Hopkins-Nanjing Center, a Sino-American academic joint venture based in China. She has taught in China and Hong Kong and served as President of Colorado College and dean at several institutions. She has written and spoken extensively about the world-class university phenomenon based on research conducted as a member of the New Century Scholars program supported by the Fulbright Program.

**Naila Nabiyeva** is Associate Professor at Azerbaijan University of Languages. She successfully completed the Junior Faculty Development Program of the Bureau of Educational and Cultural Affairs of the U.S. Department of State at Seton Hall University in the 2008–2009 fellowship year. Her scholarly interests, within the field of Education Administration, include the following: assessments of student learning, faculty instructional development, technology and its impact on teaching and learning functions, and values and ethics in educational leadership.

**Diane Brook Napier** is Associate Professor at the University of Georgia (Department of Workforce Education, Leadership, and Social Foundations), Head of the Program of Social Foundations of Education, and Member of the UGA Institute of African Studies. Her teaching and research in comparative and international education focus on issues of human resource development, educational transformation, and policy implementation in post-colonial states, particularly in sub-Saharan Africa and especially in her home country of South Africa, but also in Cuba and the United Arab Emirates. Her teaching and research interests include dimensions of deracialization of education and post-colonial race relations, identity and citizenship, language and gender rights and policy, environmental justice, teacher education, skills development, higher education transformation, and U.S. influences in the global-local educational reform continuum.

**Kim Dung Nguyen** is Vice Director General of the Institute for Educational Research, University of Pedagogy, Ho Chi Minh City, Vietnam. She is considered an expert on higher education accreditation in Vietnam and has served as a consultant to several international organizations, including the World Bank and the Vietnam-Netherlands Higher Education Project.

**Isaac Ntshoe** is an Associate Professor in the Institute of Open and Distance Learning, UNISA. His areas of research interest include higher education policy and planning, globalization, and the internationalization in higher education.

**Diane E. Oliver** is Assistant Professor in the Department of Educational Research and Administration, Kremen School of Education and Human Development, California State University, Fresno. She has authored and coauthored several published pieces and presented papers on higher education and community colleges in Vietnam, particularly with reference to the influences of globalization.

**Francisco O. Ramirez** is Professor of Education and (by courtesy) Sociology at Stanford University. His current research interests focus on the impact of world society on the content of models of the university and on the institutionalization of human rights and human rights education. His work has been supported by grants from the National Science Foundation, the Spencer Foundation, and the Freeman Spogli Institute for International

Studies at Stanford University. His most recent publications are forthcoming in *Sociology of Education* and *Social Forces.*

**Robin Shields** is Lecturer in Education Studies (Global Education) at Bath Spa University and has experience in running education programs in the international development sector. He completed his PhD in Education at the University of California, Los Angeles on the use of information technology for rural distance education in Nepal and has published several articles on educational policy and practice in Nepal.

**Ashley Shuyler** is the founder of AfricAid (www.africaid.com), a nonprofit organization that supports girls' education in Africa. She graduated from Harvard University in 2008, with a degree in Social Studies, specializing in international development. While there, she focused her research on Tanzania's national examination system and its consequences for educational and economic development. She is currently serving as the Director for AfricAid, working to develop and oversee the organization's programs in East Africa.

**Joseph Stetar** is a Professor of Education at Seton Hall University. His research interests center on international education policy and development. He has received support for his work from USAID, USIA, and the Andrew W. Mellon Foundation, among others. In 2006 he served as an Embassy Policy Specialist at the U.S. Embassy, Bishkek, Kyrgyzstan under the International Research and Exchanges Board (IREX) Program.

**Frances Vavrus** is an Associate Professor of the program in Comparative and International Development Education at the University of Minnesota. Her research and teaching focus on development education and foreign aid policy, teacher education in Sub-Saharan Africa, and gender studies. Her previous publications include her book *Desire and Decline: Schooling amid Crisis in Tanzania* and articles in *Compare, Gender and Education*, and the *Harvard Educational Review.* She also serves as Advisor to the Board of AfricAid and directs its *Teaching in Action* program for secondary school teachers in Tanzania.

**Yingjie Wang** is a Professor and Senior Scholar in Comparative and International Education at Beijing Normal University, one of China's leading institutions for teacher training and education research. In recent years it has expanded into a comprehensive research university while maintaining its historic focus on education issues. Wang has served as dean and vice president of his home institution; he has also been a Visiting Scholar at Stanford and Harvard Universities to pursue research on U.S. higher education. Recent publications relevant to the theme of this book include

"Rules and Enlightenment: Considerations on Building World-Class Universities" and "University Ranking: Problems and Suggestions."

**Gerald Wangenge-Ouma** is a Postdoctoral Fellow and Coordinator of the Master's of Education Program in Higher Education Studies at the University of the Western Cape in South Africa. His main research topics are financing higher education and higher education policy. Recent publications include articles in *Higher Education, Journal of Higher Education Policy and Management, South African Journal of Higher Education*, and *Oxford Review of Education*.

**Anthony Welch**, Professor at the University of Sydney, specializes in education policy sociology, with particular interests in Australia, East Asia, and Southeast Asia. His current work focuses largely on higher education reforms. A recent Fulbright New Century Scholar, he has also been Visiting Professor in Germany, the United States, Japan, the United Kingdom, and France. Recent books include *The Professoriate* (2005) and *Education, Change and Society* (2007). A volume on China and Southeast Asian relations is in press, with a further book on Southeast Asian higher education forthcoming. He also directs a nationally funded research project called "The Chinese Knowledge Diaspora."

# Index

CPSIA information can be obtained at www.ICGtesting.com
Printed in the USA
LVOW05s1402060114

368266LV00003B/289/P